EXOPLANETS

Exoplanets

Hidden Worlds
and the Quest for
Extraterrestrial Life

DONALD GOLDSMITH

Harvard University Press

Cambridge, Massachusetts • *London, England*

2018

First Printing

Library of Congress Cataloging-in-Publication Data

Names: Goldsmith, Donald, author.
Title: Exoplanets : hidden worlds and the quest for
extraterrestrial life / Donald Goldsmith.
Description: Cambridge, Massachusetts : Harvard University Press, 2018. |
Includes bibliographical references and index.
Identifiers: LCCN 2018005407 | ISBN 9780674976900
(hardcover : alk. paper)
Subjects: LCSH: Extrasolar planets. | Extrasolar planets—Detection. |
Life on other planets.
Classification: LCC QB820 .G64 2018 | DDC 523.2/4—dc23 LC record
available at https://lccn.loc.gov/2018005407

Book design by Chrissy Kurpeski

For Oriana and Felix—
and a new generation of exoplanet observers

CONTENTS

EXOPLANETS

PROLOGUE

n 2015, two experts deeply involved in the expanding efforts to detect and to study planets around other stars wrote that "there are few precedents in the history of science in which a discipline moves so rapidly from shaky disrepute through a golden age of discovery and into a mature field of inquiry. In less than two decades, the study of extrasolar planets has accomplished all of this—answering questions that were posed at the dawn of the scientific era, while affording tantalizing glimpses of revelations to come." Gregory Laughlin of the University of California, Santa Cruz, and Jack Lissauer of NASA's Ames Research Center have heralded these successes, but they also note that "this rapid progress partially obscures the fact that the extrasolar planets are *fundamentally alien*. Virtually none of their properties, either statistical or physical . . . were predicted or anticipated."[1] Another exoplanet expert, Scott Gaudi of Ohio State University, puts a more positive spin on this startling result: "'Mother nature is clearly more imaginative than we are.' In other words, we are continually surprised at the diversity of exoplanetary systems, and how different they are from our own solar system."[2]

The flood of exoplanet discoveries during the past two decades has continually underscored the hazards of employing slim evidence in leaping to broad conclusions. Paul Butler, one of the pioneers in this field, has reminded us that

> prior to the discovery of exoplanets, everyone from the world's leading theorists to popular science fiction culture . . . imagined that most planetary systems would look similar to the solar system, with small rocky planets orbiting in the inner few A.U., giant planets further out, all in majestic concentric circular orbits. The reality could hardly be more different. The implications of this are profound. . . . With zero examples of a phenomenon, we are forced to use imagination. With one example we are wearing a pair of blinders. All other avenues are excluded. I am sure that these "blinders" affect most of us on everything ranging from the mundane details of daily life to the search for life in the universe.[3]

Science doesn't offer opportunities any better than the challenges posed by unsuspected results from long-unanswered questions. The bounty of new evidence about the worlds that orbit other stars has resonated through the world of astronomy, creating new careers for hundreds of scientists and stimulating the creation of an array of ingenious new instruments, both Earthbound and spaceborne, that promise to advance our knowledge in ways that will make today's reams of astronomical data seem entirely modest.

Research into exoplanets (this term has won more general acceptance than "extrasolar planets," although younger readers might prefer "XOplanets") took a giant step upward with the announcements in late 1995 and early 1996 that astronomers had found a planet around each of seven stars in the sun's neighborhood.[4] The stars had characteristics similar to the sun's, but many of their planets bore only slight resemblance to objects in the solar system. Three of them are giant planets that take only a few days to complete their orbits, and they have diameters less than 6 percent of the diameter of the Earth's orbital path around the sun. Since then, as one startling discovery has followed another, astronomers have had the happy task of explaining how the exoplanets that their brilliant techniques revealed now fit into—or violate—the previous models that they had put forth.

As an astronomer and astronomy popularizer, I leapt at the chance to write the first book about the first seven exoplanets, and I almost succeeded with *Worlds Unnumbered,* published toward the end of 1996. The book's appearance coincided with Govert Schilling's *Tweeling Aarde—De Speurtocht Naar Leven in Andere Planetenstelsels.*[5] Schilling's title and idiom tended to limit its readership, while I have no convenient explanation for the limited sales of my opus.

In those exciting days, the story of how astronomers found the first exoplanets, and what they implied for our understanding of how other planetary systems formed and evolved, seemed to unfold with a majestic simplicity compared with today's more complex situation. As we approach the twenty-fifth anniversary of these discoveries, we may rightly revel in the abundance and variation among the new worlds revealed by ever-improved telescopes and spacecraft. Although this decade's astronomers count exoplanets in the thousands, in the next they will know them by the tens of thousands. To the single technique used for their initial exoplanet discoveries, astronomers have now added half a dozen. Each technique has its own advantages and disadvantages, and astronomers have now begun to employ some of them simultaneously, evoking a synergy that sharpens our knowledge. I invite my readers to join me in sailing through the Milky Way (with a nod to the vast empires of space beyond) to examine worlds both known and unknown, laden with the promise of yielding their secrets—including their suitability for extraterrestrial life—as well as the burden of offering astronomers immense challenges in their attempts to bring these secrets to light.

1

*

THE LONG SEARCH FOR OTHER SOLAR SYSTEMS

Throughout the past few millennia, well before humans understood the layout of the cosmos, philosophers (and ordinary people as well) have engaged in enjoyable speculation about the possibility of worlds beyond our own. During the first century BCE, as Julius Caesar came to dominate what soon became the Roman Empire, his contemporary Titus Lucretius Carus wrote a famous poem, *De Rerum Natura (On the Nature of Things)*, in which he demonstrated—to his own satisfaction—that "in the universe there is nothing single, nothing born unique." Lucretius drew the natural conclusion that "you must therefore confess that sky and earth and sun, moon, sea and all else that exists are not unique, but rather of number numberless."[1] Thus the cosmos must contain "worlds unnumbered," as Alexander Pope wrote 18 centuries later.[2] But does it?

The alternative view, at least equally attractive to many, placed our world at the center of the universe and regarded it as not necessarily similar to any other object. From an intuitive viewpoint, nothing could be more obvious than our cosmic centrality. We live on an apparently fixed, unmoving planet, around which we observe (at least before city lights took away the night sky) all celestial objects in ceaseless motion as they wheel across the skies by day and by night. Every society whose creation myths have survived has supported a belief system that privileges our Earth above all else, implying, or definitively stating, that cosmic forces observe, protect, and govern earthly events.

Throughout the past half-millennium, scientific and technological advances have gradually eroded the public's belief that we occupy the center of the universe. A sizable fraction of humanity no longer consciously adopts this attitude. Aware of our cosmic mediocrity, many of us have come to adopt an outlook much closer to Lucretius's—a belief that a host of worlds populate the cosmos. Most of us have absorbed the fact that our sun ranks as a near-typical star among the multibillion, star-studded throng of our Milky Way galaxy, though the details may evade us. In addition, by employing the ability to reason by extrapolation that has served humanity so well, most of these multiworlders have asserted, on entirely reasonable grounds, that the enormous numbers of stars, and the enormous numbers of the planets assumed to encircle them, imply that life must be abundant throughout the cosmos, and that at least some of these planets harbor forms of intelligent life that rival or surpass our own.

Part of this chain of reasoning has proven entirely correct: We now know that a large fraction—perhaps the majority—of the vast swarm of stars does possess planets. The flood of discoveries during the past two and a half decades has resulted in nearly 4,000 verified exoplanets, along with thousands of candidates ripe for further examination. The enormous variety of the exoplanetary horde embraces many objects whose sizes, masses, composition, and distances from their parent stars deviate markedly from the expectations induced by a natural, though in the event misdirected, tendency to surmise that our own planetary system serves as a model for others.

So far as the extrapolation to extraterrestrial life goes, we must state the obvious: We have no strong indication that any such life exists, but we recognize that the absence of evidence does not constitute evidence of absence. Life may or may not exist on many of the recently found extraterrestrial worlds, or on the far greater numbers of worlds soon to be found (based, once again, on extrapolation from our current findings). Conditions that exist on many, though not most, of these planets may well favor the origin of life as we know it. But although we have firm and abundant evidence that extrasolar planets exist, our discussions of the possibilities of life beyond Earth remain largely speculative.

In contemplating extraterrestrial life, our natural human tendencies push us toward the search for a planet most like our own, often called Earth 2.0. But if the multitude of planets found around other stars has a single strong lesson to teach us, as well as the astronomers who have fallen into a similar mental aberration, we would do well not to concentrate overmuch on this quest for Earth's twin. If, as seems reasonable, the greatest fascination that most of us experience in contemplating the worlds that populate our galaxy resides in the hope (or fear) that the beings who may exist upon them have much to teach us, then we should heed the lessons of the past and avoid restricting ourselves, as earlier speculations have often proposed, to concluding that any such beings must, or are even most likely to, appear on planets that most closely resemble our own.

This book aims to present our current, rapidly evolving knowledge of other planetary systems, which springs from at least seven different, often complementary and interlocking, discovery techniques. Astronomers' spaceborne and ground-based searches draw on hard-won understanding of how the laws of physics underlie and explain the essence of the cosmos. The basic physics behind the quest for exoplanets includes the laws that govern gravitational attraction and celestial dynamics; Einstein's general theory of relativity; the rules of optics and what they imply about limiting and improving our views of the universe; and the spectroscopic analysis of light waves and their x-ray, ultraviolet, infrared, millimeter-wave, and radio cousins.

Among the sevenfold pathway of techniques that astronomers now employ in their search for exoplanets, three approaches have provided the bulk of known exoplanets. First came measurements of how a planet's gravitational force on its star affects the star's motions, which can reveal not only the planet's existence and a lower limit on its mass, but also the size and elongation of its orbit. Next, astronomers found planets whose orbits happen to carry them across our line of sight to their stars, first with ground-based observations and then, in far greater numbers, with spaceborne observatories that can monitor stellar brightnesses with amazing precision. Third (by the number of exoplanets discovered), astronomers used another effect of planets' gravitational forces—their ability, predicted by Einstein's relativity theory, to focus and to distort the light from much more distant stars—to find planets at impressively large distances from the solar system.[3] Although exoplanets at any distance from the solar system deserve attention, those closest to us have a greater appeal. We can dream more reasonably of potential explorations of these planets (see Chapter 14). Far more important for the next few decades will be the opportunities that astronomers will have to study the exoplanets closer to us with more techniques, and with greater accuracy, simply because their proximity makes them appear brighter to us.

Subsequent chapters will examine the three chief methods for finding exoplanets, as well as four subsidiary approaches that have brought success and offer breakthroughs in the years to come. Before we examine future opportunities to locate and to understand new worlds in the cosmos, we will examine astronomers' current theories of planetary formation, which, quite understandably, have been heavily influenced by what we now know about exoplanets.

The impressive results of 25 years of exoplanet observations should soon be far surpassed by the observational fruits of an array of instruments to be created and launched during the next two decades. Each of our current techniques for finding exoplanets directs astronomers into planning for future years, when exoplanet science will surpass its current, well-earned maturity. The present and future study of exoplanets offers the joy of searching for Earth's cousins, some of which may harbor systems of living organisms whose evolution, though analogous to our own, has yielded quite different results. But even lifeless worlds have their own appeal—as discoveries within our solar system have shown—that justifies our attempts to learn as much as we can about them.

2

·

COSMIC DISTANCES

The vast distances that separate objects throughout the universe provide the most significant, and in some ways the most evident, feature of our cosmic surroundings. Most notably, the distances between the stars, and thus the distances between any planetary systems that may surround them, exceed what human intuition suggests by enormous factors. The strangeness of the universe begins with distances that surpass easy understanding.

Astronomers have now concluded that two mysterious, invisible, and entirely disparate entities—dark matter and dark energy—permeate and dominate the universe in mass and energy terms.[1] Dark matter, revealed by its gravitational effects on "ordinary matter," consists of particles that are entirely unknown to us at the present time. The "ordinary" form of matter resides primarily in vast clouds of hot gas that permeate giant clusters of galaxies; to a lesser extent, we find ordinary matter in the stars that form the visible units of the universe. The contribution from any smaller objects that orbit these stars falls far below the amount of matter in the stars themselves. Dark energy, even more mysterious, steadily increases the rate at which the universe expands. Happily for our purposes, neither dark matter nor dark energy significantly affect the search for planets that may orbit our stellar neighbors.

During the middle of the nineteenth century, as astronomers first measured the basic distance scales of the cosmos, they realized that the immense distance from the Earth to the sun (150 million kilometers) represents only a tiny fraction, about 1 part in 300,000, of the distances to the nearest stars. Less than a century later, a better-equipped generation of astronomers showed that the distance to those closest stars equals only about ½₅,₀₀₀ of the diameter of the Milky Way galaxy, the cosmic collection of several hundred billion stars within which our solar system occupies a suburban location far from the galactic center.[2]

Because the distance numbers grow so rapidly (for example, the Milky Way has a diameter roughly 6 billion times larger than the Earth–sun distance), astronomers developed new ways to specify cosmic distances. These new units, the light year and the parsec, measure the distance that light travels in one year (slightly less than 10 trillion kilometers) and the distance to an object at which the Earth's yearly motion around the sun changes its apparent location on the sky by 1 second of arc in each direction (about 31 trillion kilometers, or 3.26 light years). Armed with these units, astronomers grew more comfortable in specifying the distance to the nearest stars (4.4 light years, or 1.35 parsecs) and the diameter of the Milky Way (100,000 light years, or 31,000 parsecs). But even these enormous units of distance proved inadequate once astronomers estimated the distances to other galaxies, millions or billions of light years away. This created a need for megaparsecs and gigaparsecs—millions and billions of parsecs, respectively.

Those who confine our attention to the study of events within our own galaxy recognize that current searches for extrasolar planets take us no farther than the kiloparsecs (thousands of parsecs) that measure large distances in the Milky Way. A journey from the solar system to the galactic center would carry us across 8 kilo-parsecs, or about 26,000 light years. As we travel along most of this trajectory until we reach the far more crowded central nucleus of the Milky Way, we would find that on the average, space contains one star, or one multiple-star system, in every cubic light year. Our immediate surroundings are far more sparsely populated. The spherical region around the sun out to a distance of 4 parsecs (13 light years) contains about 2,800 cubic light years, within which astronomers have found 30 star systems: one system in every 93 cubic light years. In our immediate neighborhood, the separation between neighboring star systems equals 4 or 5 light years.[3]

A moment's reflection highlights the difficulty of finding planets around even the closest stars, which involves reflection in its literal aspect. Planets emit essentially no light of their own, though some of them do emit significant amounts of infrared—radiation with wavelengths longer than those of visible light. A planet therefore shines in visible light only because it reflects some of the light from its own star, in an amount that depends on the planet's size, reflectivity, and distance from its star. The Earth, for example, intercepts about one-billionth of the light that the sun generates and reflects about 30 percent of it into space. As a result, an astronomer on a planet in another system would "see" the Earth shining in visible light with about three 10-billionths of the sun's brightness. If the Earth shone this brightly with no sun present, finding it amid the blackness of space would not prove especially difficult for modern telescopes, but the same fact that allows the Earth to shine at all—its comparative proximity to the sun—also hides it within the sun's much greater glare, making it nearly impossible to detect. Astronomers like to compare this type of task to the attempt to find a firefly next to a searchlight, though in fact the task is astronomically more difficult, and more like trying to find a bug with one-billionth of the brightness of a firefly in the searchlight's glare.

Although planets do not produce visible light, most of them emit some infrared radiation from their internal heat. This infrared glow offers the chance for direct observation of the largest giant planets at relatively great distances from their stars, because in infrared radiation, the star outshines the planet by only a few million times, instead of the billion or so times in visible light.

The searchlight-firefly problem had long convinced astronomers that their attempts to find extrasolar planets should rely on indirect methods, which would not detect the planets themselves but their *effects* on the stars around which they orbit. The vast majority of early exoplanet detections employed such indirect methods, which—with important exceptions—will continue to provide a key detection method in the decades to come.

3

·

EARLY QUESTS
FOR EXOPLANETS

The diverse discovery techniques that astronomers use in their searches for exoplanets embody a reflection of those researchers' imaginations, insights, and determined efforts. In a review of the history of methods used in the quest to find other worlds, Virginia Trimble, an astronomer at the University of California, Irvine, listed two dozen possible approaches, some already tested through use, but most of them not yet undertaken, for obvious reasons. For example, the final two entries in Trimble's list are (a) the arrival of extraterrestrial visitors and (b) "something even more outlandish."[1] For our purposes, however, we need consider only the top seven or eight methods of finding planets; their characteristics and results appear in the three exoplanet catalogs cited in the Further Reading section of this book. Most of the nearly 4,000 verified exoplanets have been found by the two major search techniques, the radial-velocity and transit methods, both of which find planets through the close observation of their stars.

Measuring the Motions of Stars with Precision

Understanding astronomers' approaches to finding exoplanets begins by contrasting two related approaches: *astrometry,* which has so far produced few positive results, and the *radial-velocity* method, which has opened the gates of exoplanetary research.[2] Both of these techniques rely on the fact, first demonstrated by Isaac Newton, that whenever a less massive object orbits a more massive one, both objects actually move in orbit around their common center of mass.[3] This center of mass lies along the imaginary line that connects the objects' centers, and the ratio of the objects' distances from the center of mass equals the *inverse* ratio of the objects' masses. Thus, for example, because the moon has $1/81$ of the Earth's mass, the center of mass of the Earth–moon system lies along the Earth–moon line, at $1/81$ of the distance from the center of the Earth to the center of the moon. This puts the center of mass inside the Earth, though closer to the Earth's surface than to its center. Each month, as the moon follows its orbit around the Earth—or, more precisely, around the center of mass of the Earth–moon system—the Earth likewise, in perfect synchrony, moves in its own orbit, $1/81$ the size of the moon's, around that center, always on the opposite side of that center from the moon.

If we expand our view to imagine a simplified version of the solar system that consists only of the sun and Jupiter, we can see that because Jupiter has $1/1,047$ of the sun's mass, the center of mass of the sun–Jupiter system must lie along the sun–Jupiter line, at $1/1,047$ of the distance from the sun's center to Jupiter's center. Since the sun–Jupiter distance equals about 778 million kilometers, the center of mass's distance from the sun's center equals about $1/1,047$ of this distance, or 743,000 kilometers, which places it just outside the sun, whose radius equals 695,000 kilometers. While Jupiter orbits the center of mass once every 12 years, the sun follows its own 12-year orbit, smaller by the factor of 1047. The existence of the other planets, along with their moons, and the asteroids, comets, and other debris in orbit around the sun, complicates the situation but does not change its fundamental character, because Jupiter has far more mass than the total mass of all the other orbiting objects.

An observer in a nearby planetary system who studied the sun would have two methods to deduce the existence of Jupiter despite being unable to see the planet directly. Both of these techniques depend on the slight wobble that Jupiter produces in the sun's motion as the solar system orbits the center of the Milky Way galaxy.

The first approach, astrometry, searches for deviations from a straight line in the sun's apparent motion across the sky of the observer. The stars in the Milky Way typically move in giant, nearly circular orbits around the galactic center, some 26,000 light years away, with each star's orbit slightly different from all others. Over timescales measured in mere years or decades, these paths essentially amount to straight lines. The slight differences among the stars' orbital motions give each star, when observed from a distance, its own "proper motion," to use an astronomical term of art. Jupiter's gravitational force would superimpose a sinusoidal dance onto the sun's proper motion, making it diverge to one side

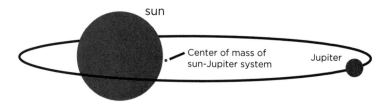

Figure 1 This highly exaggerated diagram of the sun–Jupiter system represents the sun, which has 10 times Jupiter's diameter and 1,047 times its mass, together with the system's center of mass, which lies just outside the sun. In reality, Jupiter's distance from the sun exceeds the sun's diameter by more than 500 times.

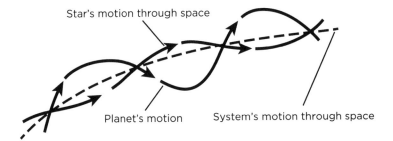

Star's motion through space

Planet's motion   System's motion through space

Figure 2 This exaggerated diagram compares the lesser motion of a star and the greater motion of its planet as the star–planet system moves through space. In reality, a planet like Jupiter would follow a sinusoidal path about 1,000 times larger than the star's. The two objects would always lie on opposite sides of the path of the motion of the system's center of mass.

and then to the other of its otherwise straight path. As Jupiter orbits the sun, the sun's dance, coincident in time with the planet's 12-year period, could be revealed by the astrometric method that measures the sun's trajectory precisely.

The second approach, the radial-velocity method, would search for a complementary 12-year dance that the sun would perform in the direction perpendicular to the plane of the sky. An astronomer from another world could try to detect changes in the sun's "radial velocity"—its velocity *along* the line of sight from the observer to the sun, rather than *across* the sky that the astrometric technique seeks to measure.

Astronomers can measure these radial velocities through spectroscopic analysis, which studies the component colors of the "spectrum" of the light from a star—in scientific terms, the component frequencies and wavelengths of starlight, which specify the precise color that our eyes may detect. The relative motion of a star toward or away from an observer produces changes in these frequencies and wavelengths, and the sizes of these changes vary in proportion to the amount of the relative motion along the

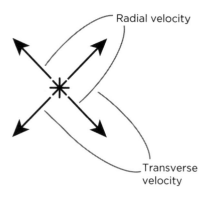

Radial velocity

Transverse velocity

Earth

Figure 3 The motion of a star with respect to the Earth can be divided into two components: "radial velocity," the star's motion toward or away from the Earth, and "transverse velocity," the star's motion across our line of sight.

observer's line of sight to the star. The phrase "relative motion" emphasizes that it makes no difference whether the star or the observer (or both) moves; all that counts is the rate at which the distance between the star and the observer increases or decreases. Astronomers, who have acquired a detailed knowledge of the complex spectra of the light from different types of stars, have learned how to compare the spectra of any starlight that they observe with the spectrum that would appear if no such motion existed. The shifts in frequency and wavelengths that arise from the stars' radial velocities, deduced from studies of the stars' spectra, directly indicate the amounts of these velocities.

As stars orbit the center of our Milky Way galaxy, their motions include random components that produce nonzero radial velocities between them, typically amounting to a few to a few dozen kilometers per second. The radial-velocity method of exoplanet detection seeks to find slight increases and decreases in these radial velocities that arise from the gravitational pull of an exoplanet in orbit. For example, Jupiter's gravitational force, exerted from different points of its orbit around the sun, would alternately increase and decrease the sun's radial velocity, and these changes would repeat themselves over a 12-year period that would mirror the dance of the astrometric deviations.[4]

An observer seeking to find Jupiter by either method would confront some basic astronomical facts. The astrometric approach would require the detection of tiny divergences of the sun's motion from a straight-line trajectory across the sky. For an observer on a planet orbiting one of the three stars in the Alpha Centauri star system, which includes the sun's closest neighbors, the observed angular separation between the sun and Jupiter would reach a maximum value of about 3.9 seconds of arc. As Jupiter orbits the sun, the planet would deviate by this angle first to one side and then to the other side of the sun's path, while the sun would deviate from its straight-line motion by only $1/1,047$ of this amount. Detecting such tiny excursions from the averaged, straight-line motion posed, and still poses, an enormous challenge to Earth-based astronomers.

The radial-velocity technique faces a similar problem. Because a planet and its star take the same amount of time to orbit their center of mass, and because the ratio of the sizes of the two objects' orbits equals the inverse ratio of their masses, the ratio of the speeds with which the star and planet move in their orbits must also equal the inverse ratio of their masses. As Jupiter orbits the center of mass of the solar system at 13 kilometers per second, the sun performs its own orbit, smaller by a factor of 1,047, and therefore moves at about 12.4 *meters* per second, close to 28 miles per hour. Jupiter would therefore superimpose continuous increases and decreases on the sun's average radial velocity, which would repeat themselves in a sinusoidal fashion over a 12-year period. In the radial-velocity approach, as with the astrometric approach, the goal of finding exoplanets must confront the fact that such radial-velocity variations rank among the smallest that astronomers can measure. For many years, detecting radial-velocity changes of a few meters per second remained beyond astronomers' capabilities.

The first claim of the detection of planets around another star arose from use of the astrometric method. Fifty years ago, after twenty-five years of studying Barnard's star, the sun's closest stellar neighbor beyond the Alpha Centauri system, the astronomer Peter van de Kamp announced that he had found small deviations in the star's path through space that implied the existence of several planets with masses similar to Jupiter's in orbit around the star. Further observations, however, found that van de Kamp had erred in his observations, and by 1973, the supposed planets around Barnard's star had vanished into history's dustbin.[5] (Stay tuned: Better methods may yet reveal that Barnard's star has planets— but they will not be the ones that van de Kamp announced.)

Today, astronomers can cite only a single, partial success in their attempts to discover an extrasolar planet by astrometry. Observations of the double-star system known as HD 176051 have found a planet, with a mass similar to Jupiter's, moving in a three-year orbit around one of the two stars—but which star remains a mystery. The large-scale astrometric discovery of exoplanets poses extreme difficulties for astronomers using current ground-based telescopes, because the ever-rippling atmosphere makes images dance, in what the public calls the "twinkling" of the stars. This renders astronomers' attempts to measure stars' changing positions with the accuracy required to reveal the existence of planets around them a near impossibility. Finding exoplanets through astrometry will require either new instruments for ground-based observatories, or the use of spacecraft free from the blurring effects of the Earth's atmosphere. These latter efforts began in 2014 with the Gaia spacecraft described in Chapter 13, and they may soon produce numerous positive results.[6]

Finding a Solar-System Planet by Its Gravitational Effects

In recent years, astronomers who analyze orbits of objects in our solar system may have found a planet in the system's far outer reaches, at a distance hundreds of times farther from the sun than any of the now-familiar "gang of eight." The hypothetical "Planet Nine"—deserving of its name if its existence should receive verification—drew attention because of its alleged gravitational effects on "trans-Neptunian objects" in the solar system. (The objects' name, and the discovery technique, remind us of 1846, when astronomers discovered the planet Neptune by observing its gravitational effects on the orbit of Uranus.) As their designation implies, trans-Neptunian objects orbit at distances from the sun far beyond its outermost planet, Neptune. Astronomers describe solar-system distances in terms of "astronomical units" (AU), each of which equals the average Earth–sun distance. We can specify Neptune's distance from the sun as 30 AU, and we should note that some trans-Neptunian objects have orbits that carry them out to more than 1,000 AU from the sun.

Sedna, one of the largest and most unusual of these objects, has almost the same size as Ceres, the sun's largest asteroid, which orbits the sun between Mars and Jupiter, but Sedna follows a highly elongated, 11,000-year-long orbit that carries it from 76 AU at its closest approach to the sun out to 936 AU at its farthest. In 2014, astronomers recognized that about a dozen trans-Neptunian objects with roughly similar orbits, often called "sednoids," have orbital characteristics that closely match one another. Because gravitational perturbations from the sun's four giant planets (Jupiter, Saturn, Uranus, and Neptune) tend to produce random orbital characteristics, some astronomers have proposed that the orbital similarities observed in the sednoids must arise from the existence of a sizable planet with an orbit even larger than the extreme distances attained by the sednoids.[7]

In 2016, astronomer Michael Brown, who is best known as the man who slew Pluto as a planet (his Twitter handle is @PlutoKiller), made a more detailed analysis of these orbits in collaboration with his Caltech planetary-science colleague Konstantin Batygin.[8] The astronomers concluded that half of the sednoids' orbits might show only gravitational influences from Neptune, but six of them, including Sedna itself, display a remarkable similarity: They all orbit in nearly the same plane and make their closest approaches to the sun in the same direction. Brown and Batygin judged the probability of this happening by chance at less than 1 percent. Instead, their analysis suggested the existence of a planet with about 10 times the Earth's mass, moving in an elongated orbit at distances from the sun that vary from 200 and 1,200 times AU, with its orbital plane highly inclined to the plane of the sun's inner eight planets. The gravitational effects from this "Planet Nine" would have corralled the six sednoids' orbits into their current similarity.

On the one hand, if Planet Nine obeys these predictions, it most likely occupies one of the farther points of its orbit, because objects with highly elongated orbits spend most of their travels near their greatest distances from the sun, where they move most slowly. At a distance from the sun 40 times Neptune's and 1,200 times Earth's, Planet Nine could hardly be said to course through interstellar space, since the sun's closest neighbor stars are 270,000

AU away, but its enormous distances from the sun would still leave it languishing in extreme cold. Frozen or not, its existence would heighten the mysteries of how the solar system—and by implication other planetary systems throughout the Milky Way—came to form with their current characteristics (see Chapter 11). We can draw one straightforward conclusion from the possibility of such a planet in our solar system. Despite astronomers' recent tremendous successes, we remain far from having the ability to find an exoplanet moving in anything similar to Planet Nine's orbit around another star.

On the other hand, of which astronomers possess a goodly number in their manifold arguments, Planet Nine may not exist at all. After completing a four-year telescopic survey of the outer solar system that revealed more than 800 trans-Plutonian objects, Cory Shankman of the University of Victoria and his colleagues concluded that the alignment of sednoids' orbits cited by Brown and Batygin disappears after detailed statistical analysis.[9] Unsurprisingly, Brown and Batygin dispute this conclusion and assert that Shankman and colleagues' analysis offers only neutral results, or in fact adds strength to the argument for Planet Nine.[10] Our look to future, deeper searches for the sun's potential outermost planet offers a timely nod to how science makes progress as we wait to see which astronomical argument better corresponds with reality.

The Pulsar Planets: An Astronomical Anomaly

History requires another modest digression before we plunge into the new depths of exoplanet research: Astronomers made the first confirmed detection of extrasolar planets not in 1995 but in 1992, and not with the astrometric or the radial-velocity method but with *radio* observations made at the Arecibo Observatory in

Puerto Rico. Although the astronomical community honors this feat, the product of hard work by Aleksander Wolszczan and Dale Frail, these extrasolar planets have failed to capture much attention in the wider world—for an understandable reason.[11]

These planets orbit not an ordinary star but a "neutron star," the collapsed core of a star that has undergone a supernova explosion and should therefore be more properly known as a "neutron core." A small minority of stars do not end their lives as faint, slowly cooling white dwarfs, the fate that lies in store for our sun about six billion years in the future. Instead they explode violently, and their explosions produce different outcomes. In some cases, the explosion blows the star completely apart; in others, the star's central region collapses so violently that it becomes a black hole. Still other explosions leave behind not a black hole but an object made almost entirely of neutrons, the result of the breakup of nuclei during the core's collapse and the subsequent squeezing together of its protons and electrons to form neutrons. A typical neutron star packs a mass roughly equal to the sun's into a volume the size of Manhattan, and therefore it has a density many trillion times the Earth's average density. Following the laws of physics, the squeezing of the star's mass into a city-sized volume concentrates the star's magnetic field and also increases its rotation rate, by enormous factors. Neutron stars typically possess magnetic fields billions of times stronger than the sun's, and they rotate extremely rapidly, in some cases hundreds of times each second.

The neutron star's immense magnetic fields and rapid rotation transform it into a giant spinning magnet, which accelerates charged particles near its surface to speeds that approach the speed of light—this is a natural, super-sized example of the conditions that physicists now create inside particle accelerators. In both cases, charged particles moving at near-light velocities in the presence of strong magnetic fields inevitably produce "synchrotron radiation," or streams of photons, the particles that form light and all other types of electromagnetic radiation. (This form of radiation was named after the synchrotron, an early type of particle accelerator.) The types of photons created by this process depend on the strength of the neutron star's magnetic field and the rapidity of its rotation. Typically, the photons from the vicinity of neutron stars are radio waves, which have the lowest frequencies and longest wavelengths of all photon types. Both the strength and synchrony of the radio photons produced in the area of a neutron star vary in accord with the neutron star's rotation.

The rotation of a collapsed stellar core often produces a pulsar, a source of radio waves that reach us in amounts that repeat in a cyclic fashion, with a period that mirrors the rotation period of the neutron star. This variation in the radio emission arises from the fact that the magnetic field surrounding the neutron star itself varies in strength. First recognized in 1967, pulsars have now been detected throughout the Milky Way and in neighboring galaxies as well. Their periodic variations can be measured to better than one part in a billion. Astronomers can even detect the gradual slowing of a pulsar's rotation, which occurs as the production of synchrotron radiation robs the rotating neutron star of a bit of its energy, moment by moment.[12]

Few astronomers imagined, however, that these exquisitely fine timings would reveal the presence of one or more planets. They applied logical reasoning: A supernova explosion should, in theory, destroy any planets around the star that undergoes such a titanic outburst. During its few seconds of explosive release, a supernova produces nearly as much energy as the star emitted during the millions or billions of years of its prior existence. Nevertheless, in the case of one particular pulsar, with the attractive name of P1257+12 (which describes its astronomical coordinates), Wolszczan and Frail found not one but two planets—and eventually one more—for a total of three.

P1257 + 12 belongs to the class of pulsars called "millisecond pulsars," the most rapidly rotating of the bunch, with rotation periods measured in thousandths of a second. Through careful timing of the pulses, the astronomers found variations that they ascribed to the radial-velocity changes induced by the gravitational forces from the planets around the neutron star. These three planets, originally known as PSR 1257 + 12 A, B, and C, now have names given by the International Astronomical Union: Draugr, Poltergeist, and Phobetor. They orbit the neutron star at 0.19, 0.36, and 0.46 AU, respectively, with orbital periods of 25, 67, and 98 days, respectively, and they have masses equal to 0.02, 4.3, and 3.9 times the Earth's mass, equally respectively. The three planets very roughly mimic the sun's inner three planets in their masses and orbital sizes, with a tiny "Mercury" and a super-massive "Venus" and "Earth," all significantly closer to the pulsar than the sun's innermost planets are to the sun.

Astronomers asked, not for the first or last time, who ordered that? Who could have imagined a neutron star with planets in orbit nearby? Holding to their conclusion that a supernova explosion would destroy any planets closer to it than the Earth–sun distance, astronomers concluded that these planets must have formed *after* the supernova explosion, from debris left behind in its general vicinity.

Astronomers may take some consolation in their failure to predict the existence of pulsar planets because of their extreme rarity. Although they have now found and characterized more than 2,000 pulsars, of which about 200 qualify as millisecond pulsars, astronomers have definitively established that just one other pulsar possesses a planet. This pulsar, PSR B1620–26, has a companion, probably a white dwarf; in addition, a planet moves around both of these objects, at an orbit much more distant from them, and it is far more massive than any of the planets around PSR 1257 + 12.[13] Discovered in 2003, and labeled as PSR B1620–26b (we shall meet this "b" designation throughout this book), this second planet has a mass about 10 times that of Jupiter's and a distance from the pulsar a bit greater than Uranus's distance from the sun. Thus it would take about 100 years for the planet to complete one orbit. This planet's impressive agglomeration of material may have also formed from the ashes of a supernova explosion, but its distance from the pulsar lends support to the hypothesis that the pulsar captured the planet during a close encounter with another star's planetary system.[14]

One can easily see why, even before the first true exoplanet had been found, "pulsar planets" failed to excite either those who searched for exoplanets around other stars, or those who specialized in creating models of planetary systems that even vaguely resembled the sun's. First, we should note that although twenty-five years have passed since radio astronomers found them, the four pulsar planets mentioned here are the only fully confirmed examples of this class. Second, and far more important, the search for extrasolar planets understandably concentrates on planets around ordinary stars—that is, on planets that presumably formed along with their stars, rather than from some of the debris accumulated in the aftermath of a stellar explosion. In their decision to search primarily for planets around normal stars, astronomers seem justified in letting their own situation as inhabitants of a planet orbiting a representative star guide them toward their targets.

4

.

THE BREAKTHROUGH: MEASURING RADIAL VELOCITY PRECISELY

Astronomers' first discoveries of exoplanets around sunlike stars came from observations that could measure stars' radial velocities with a precision of a few meters per second. The simplified model of the solar system outlined in the previous chapter has shown that the radial-velocity detection of a Jupiter-mass planet, orbiting a sunlike star at a Jupiter-like distance, depends on our ability to measure changes of 12 meters per second or less. For many years, this requirement posed an apparently insuperable barrier. In astronomical terms, a radial velocity of 12 meters per second—not all that fast in our terrestrial experience—represents an impressively slow speed.

There's another problem. We see the full amount by which a star's radial velocity changes only if the orbits of the star and planet happen to lie directly along our line of sight. If the plane containing these orbits tilts with respect to our line of sight, as will occur in most cases, then we will observe only a fraction of the true radial-velocity changes. If the orbital plane happens to lie perpendicular to our line of sight, none of these changes will be detectable by an observer on Earth. If we take into account all the possible, presumably random orientations of these orbital planes, the average radial-velocity changes that we observe amount to about 64 percent of those that we would see if the alignment were perfect for our detection purposes.

Newton's laws and geometry dictate that the radial-velocity method will reveal most easily exoplanets with (a) larger masses, (b) lesser distances from their stars, and (c) an orbital orientation that places the orbits almost directly across our line of sight. Among these three factors, (a) and (b), which increase the sizes of the radial-velocity changes, turn out to be more important than (c), which, on average, reduces the average radial-velocity changes from their full potential by about a third. In contrast, a planet 10 times more massive than another at the same distance will produce radial-velocity changes 10 times greater, and a planet at one-tenth the distance of another with the same mass will induce radial-velocity changes about three times greater. (The planet's orbital velocity, and the star's as well, varies in proportion to the square root of the planet–star distance.) In short, the radial-velocity method strongly favors finding massive planets close to their stars.

For many years, astronomers searching for extrasolar planets with radial-velocity measurements sought to measure velocity changes measured in tens of meters per second. Their efforts involved improvements in spectrographs, the fundamental astronomical tools that measure the amounts of different colors from a source of light with extreme accuracy. Spectrographic observations split the light from a star, planet, or galaxy into its component colors so finely that traditional color names no longer provide a useful marker. Instead, astronomers specify the wavelength or frequency at each point within the spectrum to a precision far better than 1 part in 1,000.

As we described in Chapter 3, the colors of light received from a source in motion differ from those observed from a stationary source. Any motion across our line of sight produces almost no change, but motion in the radial direction, along our line of sight, follows the simple rule first discerned by the Austrian physicist Johann Doppler. Except for speeds close to the speed of light, the fractional change in frequency or wavelength equals the ratio of the speed of the source of light to the speed of light itself, which is close to 300,000 kilometers per second. Motion away from the observer increases the wavelengths and decreases the frequencies by this ratio; motion toward the observer, in complementary fashion, decreases the wavelengths and increases the frequencies by the same ratio. The effects of these radial motions do not depend on whether the source moves, the observer moves, or both move: All that counts is the velocity with which the objects approach or recede from one another.

A spectroscopic system that can measure changes in frequencies and wavelengths by 1 part in 10,000 can detect, at the extreme limit of its capabilities, radial-velocity changes close to 30 kilometers per second. If the system can find changes by one part in a million, velocity changes of 300 meters per second may become detectable. Finding changes in radial velocity at the level of 30 meters per second requires a system operating at the level of 1 part per 10 million. And even this system could not find a Jupiter orbiting a sunlike star at a distance comparable to Jupiter's distance from the sun![1]

Two additional problems confronted astronomers who sought to employ the radial-velocity method to find exoplanets. First, the Earth's atmosphere contains a host of molecules that superimpose their own features on the stellar spectra under scrutiny, adding to the difficulty of measuring tiny changes in the observed frequencies and wavelengths of the starlight. Second, stars stubbornly refuse to maintain complete constancy in their production of light. Their surfaces heave and ripple, producing radial-velocity changes of differing amounts, at different times, and in different regions of the stellar surface. Stars often develop temporary darker regions (known as sunspots in the case of our own star), and they also produce stellar flares, sudden outbursts whose extra light rises and dims on a timescale of hours or days. As a star rotates, the light arriving from different portions of its disk has different radial velocities. In the case of constant amounts of light, astronomers can take this into account fairly easily as they examine light from the entire star, but changing amounts of light from different regions on a star's surface confuse the attempts to measure the star's radial velocity with extreme precision.

For years, these physical facts of life bedeviled planet-hunting astronomers. The radial-velocity method offered great promise for finding planets through the planets' gravitational effects on their stars, with a time-tested method that demanded "only" improved equipment with a wonderful sensitivity. These improvements required decades of research and development before they finally arrived in 1995, opening our view of other planetary systems.

In examining the long history of these undertakings, we should salute the Canadian pioneers who improved the radial-velocity method, Gordon Walker and Bruce Campbell. Their labors at the Dominion Astrophysical Observatory in Canada through the 1970s and 1980s yielded a single exoplanet candidate, in orbit around the star Gamma Cephei, whose existence was seriously doubted during the 1990s; improved observational techniques finally confirmed its existence in 2003.[2] After spending decades in what seemed a fruitless search, Walker and Campbell adopted other avenues for their activities—hardly the first or last time that the first scientists to seek new worlds have found frustration. But "Gordon Walker and Bruce Campbell were the true pioneers [of exoplanet discovery]," says Alan Boss, an astronomer at the Carnegie Institution of Washington, and they should not be forgotten amidst the flood of later triumphs of the radial-velocity approach.[3]

The First True Exoplanets

In the fall of 1995, exoplanets burst upon the world with the announcement that the Swiss astronomers Michel Mayor and Didier Queloz had used the radial-velocity method to find a massive planet around the star 51 Pegasi. Within a few months, the American astronomers Geoff Marcy and Paul Butler, working at the Lick Observatory in California, had found planets around six more stars. As we look back on this discovery from the present era in which astronomers have found thousands of exoplanets, we should pause to note how thoroughly this news then gripped the public by demonstrating that our solar system, once considered as possibly unique in the Milky Way, had acquired a host of rough analogs. Astronomers cited their "Copernican principle" (named after the sixteenth-century astronomer who posited that the Earth does not occupy the center of the cosmos), which asserts the useful, often verified hypothesis that our Earth, our star, and our planetary system do not occupy a unique position in the universe.[4]

As astronomers welcomed these exoplanets, they recognized that they posed an immense problem—and a tremendous opportunity. Because all of the newfound planets shared characteristics so different from those that our solar system had suggested should exist, astronomers called for a wholesale revision in our understanding of the faraway worlds that orbit other stars.

Figure 4 shows the graph that plots the effect of the first of these exoplanets on its star. This planet, in orbit around the sunlike star 51 Pegasi and located 50 light years from Earth, carries the designation 51 Pegasi b, or 51 Peg b, for short. In naming exoplanets, astronomers soon chose a system with lowercase letters that follow their stars' names, beginning with the letter b and rejecting science fiction's numerical system, which had yielded planets such as Altair IV ("Forbidden Planet") and Omicron Persei 8

("Futurama"). Astronomers have now found planetary systems whose designations range from b through i (and may yet reach the end of the alphabet).

Radial-velocity measurements revealed that the planet around 51 Pegasi takes only 4.23 days for each orbit, implying that the planet has a distance of only 0.05 AU from its star. The Swiss astronomers' radial-velocity observations also showed that 51 Peg b has a mass at least 47 percent of Jupiter's mass. Because we do not know the angle by which the planet's orbit inclines to our line of sight, this mass, like any other mass derived for an exoplanet revealed by the radial-velocity technique, represents the *minimum*

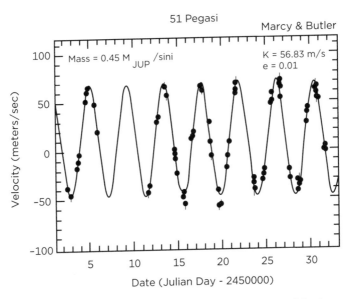

Figure 4 Radial-velocity measurements (in meters per second) of the star 51 Pegasi show the velocity at each point along the phase of the velocity changes. The vertical bars show the estimated errors in the measurements. The nearly perfect sinusoidal shape of the best-fit line through the velocity changes shows that the planet moves in a nearly circular orbit. This star's changes in radial velocity, which reach plus or minus 50 meters per second, far exceed astronomers' current best efforts in measuring such changes, which have surpassed 1 meter per second. (Courtesy of R. Paul Butler)

possible mass for the planet. Only if the planet's orbit lies directly along our line of sight would the mass deduced from radial-velocity measurements equal the planet's actual mass.

How could a planet with a mass probably less than Jupiter's produce detectable radial-velocity changes? The answer resides in the exoplanet's small distance—only 7.5 million kilometers—from its star, which allows it to induce radial-velocity deviations from the average value by 56 meters per second, more than four times the changes that Jupiter produces on the sun. The next six exoplanets proved to have minimum masses between one-half and 6.6 times Jupiter's. Like 51 Peg b, four of them have distances from their stars that are much less than the Earth–sun distance—in the case of Tau Boötis b, only 0.046 AU. The other two, however, orbit their stars at distances of 1.7 and 2.1 AU. (Useful data on these and other exoplanets appear in the websites listed in the Further Reading section at the end of this book.) Marcy and Butler went on to find 70 of the first 100 known exoplanets, while carefully crediting Marcy's mentor at Lick Observatory, Steven Vogt, for his role in creating the spectrographic system that allowed measurements of the stars' radial-velocity variations with an accuracy sufficient to reveal the exoplanets' existence.

During the final half decade of the second millennium, astronomers refined their radial-velocity observations to the point that they could detect exoplanets orbiting at distances as large as twice the Earth–sun distance, provided that the planets had masses considerably larger than Jupiter's. As the number of verified exoplanets approached 100, planet hunters, who had once proudly proclaimed that we now knew more planets outside the solar system than within it, relegated this boast to ancient history and continued to refine their analysis of their radial-velocity observations, teasing additional information about exoplanet orbits from the planets' measured effects on their stars.

Deducing an Exoplanet's Orbital Description

Radial-velocity measurements allow astronomers to derive a surprising wealth of data from the changes in a star's radial velocity measured through many planetary orbits. First, the interval between repetitions in the star's maximum or minimum radial velocity reveals the planet's orbital period. Second, subtle aspects of the changes in radial velocity show whether the star follows a circular orbit, in which case these changes follow a symmetric, sinusoidal pace, or whether the orbit shows a measurable elongation, which reveals itself by a pattern of more rapid and less rapid changes during each orbital cycle. The laws of orbital dynamics require that the *shape* of the star's orbit exactly matches that of its planet, even though the planet's orbit may be thousands of times larger.

A perfectly circular orbit will lead to a perfectly symmetric sine-wave pattern (see Figures 4 and 6), while an elongated orbit will generate a pattern with alternating regions of bunching and

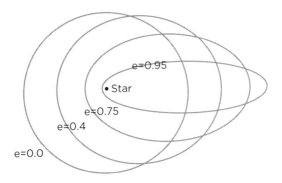

Figure 5 The shapes of elliptical trajectories are specified by their elongations, technically called "eccentricities." Each ellipse has a long axis, denoted by *a*, and a short axis, denoted by *b*. The orbital eccentricity equals the square root of the quantity $[1-(b^2/a^2)]$. If $b=a$, the orbit is circular, and its eccentricity is zero. Smaller values of the ratio (b/a) imply larger eccentricities.

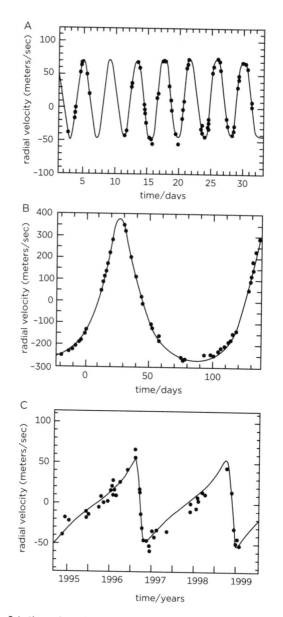

Figure 6 In these three diagrams, each dot represents a radial-velocity measurement. The shapes of these curves reveal the eccentricity of each star's orbit, which matches the eccentricity of its planet's orbit. In the first diagram, the eccentricity equals 0; in the second, the eccentricity is close to 0.3; and in the third, the eccentricity equals 0.6.

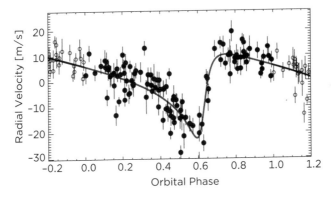

Figure 7 The changes in the radial velocity of the star HD 3651, observed through many orbits of the star and its planet around their center of mass, show a pattern noticeably different from the sinusoidal shape that would characterize a circular orbit. The vertical lines show the estimated possible error of each radial-velocity measurement. In this case, astronomers could deduce that the orbits of the star and planet are strongly elongated, with an eccentricity of 0.6. Compare this diagram with the three radial-velocity plots shown in Figure 6. (Courtesy of Debra Fischer)

spreading through each orbit (see Figures 6 and 7). The amount of the elongation of the star's orbit, which astronomers denote as the orbital eccentricity, exactly equals the eccentricity of the planet's much larger orbit, whose shape and orbital period must match the star's.[1]

In addition to furnishing us with the planet's orbital period and eccentricity, radial-velocity observations provide a third key property of the planet's orbit: its mass. As Johannes Kepler realized four centuries ago from his studies of the planets in the solar system, and as Isaac Newton showed how to generalize to other objects in orbit, the length of time required for a less massive object to complete an orbit around a more massive one depends on the average distance between the two objects. More distant objects, which feel less gravitational force, move more slowly in orbit and must travel farther to complete an orbit. Kepler's mathematical rule relates the planet's orbital period, P, to the average planet–star distance, which astronomers call a: P varies in proportion to the $\frac{3}{2}$ power of a. Or, if you prefer, using the seeming magic of algebra, a varies in proportion to the $\frac{2}{3}$ power of P. Jupiter, for instance, has about five times the Earth's distance from the sun, and it takes 5 to the $\frac{3}{2}$ power, or about 12 years, to orbit the sun.

In order to apply this relationship to another planetary system, we must know the mass of the star responsible for an exoplanet's motion. A star more massive than the sun will exert greater gravitational force than the sun does on Earth upon a planet orbiting at the same distance, which will make the planet move more rapidly. (The *planet's* mass, however, affects neither its speed in orbit nor the time required to complete each orbit.) Happily, astronomers have long since acquired a formidable ability to deduce the mass of a star from the details in its spectrum, and they can specify the mass of any normal star to an accuracy of 10 percent or less. They can therefore add a third key parameter to the planet's orbital properties by deriving from their observations an exoplanet's orbital period, its orbital eccentricity, and its average distance from its star, all with impressive precision.

Finally, if we apply the important caveat just mentioned, we can use the radial-velocity measurements to deduce the planet's mass. The planet's average distance from its star reveals the total distance around its orbit. When we divide this total by the planet's orbital period, we obtain the planet's average speed in orbit. Newton's laws imply that the ratio of the planet's velocity to the star's velocity, as it moves along its much smaller orbit, equals the *inverse* ratio of the planet's mass to the star's mass. Since we observe the star's velocity, have estimated the star's mass, and have derived the planet's velocity, we can quickly calculate the planet's mass—but only if we know that we are observing, in effect, all of the star's velocity because its motion occurs exactly along our line of sight.

And there's the problem: We have no way, in most cases, of determining the amount by which the plane that contains the orbits of the star and planet inclines to our line of sight. This has nothing to do with the orbit's intrinsic properties—only with how chance has allowed us to observe the star's motion in its tiny orbit. If we see the orbit edge-on, we observe the full extent of the star's motion in response to the planet's gravitational force. In that case, the mass that we deduce for the planet equals the actual planetary mass. If, however, the orbital plane tilts with respect to our line of sight, we observe only a portion of the star's actual velocity, namely, its actual velocity times the sine of the angle by which the orbital plane inclines to our line of sight. (An inclination of 90 degrees occurs when the orbital plane coincides with our line of sight to the star–planet system, whereas a zero-degree inclination describes an orbital plane perpendicular to our line of sight. Trigonometry buffs will recall that the sine of zero degrees equals zero, and that the sine of larger angles increases to one-half for an angle of 30 degrees, and to unity for an angle of 90 degrees.) The average value for the sine of the inclination angle, taken over all random orientations of stars' orbital planes with respect to our line of sight, equals $2/\pi$, or about 0.64, so we can state that on the average, the true masses of the exoplanets found by the radial-velocity technique equal $1/0.64$, or about 1.56 times, the masses implied by the observed changes in the stars' radial velocities. But we cannot know which planets have the same mass as the mass derived from our observations, and which have two, three, or in a few cases even more times this mass. Furthermore, we are more likely to discover planets whose orbital planes nearly coincide with our line of sight than planets with the same mass whose orbital planes are close to perpendicular to our line of sight, because the former induce larger, hence more easily detectable, changes in their stars' radial velocities. We will meet the fortunate exceptions to the uncertainty in exoplanet masses in the next chapter.

The First Exoplanet Discoveries
Pose a New Problem

As the catalog of exoplanets swelled beyond 100, astronomers grappled with the fundamental problem raised by many of these objects: How could Jupiter-mass planets come to be orbiting so close to their stars? All theories of planet formation—based on our understanding of physics and naturally referenced by the only previously known example of a planetary system—envisioned a model in which small, rocky planets orbit relatively close to their star, with much larger and more massive gas giants orbiting at far greater distances. This made excellent sense in view of the fact that as a star begins to shine, it heats its surroundings more strongly in regions closer to it. Within a certain distance, the star will drive away all of the gas nearby, so that any planets that form must be solid, as are the sun's four inner planets. In contrast, the sun's four giant planets apparently began with rocky cores that quickly attracted vast amounts of gas, a process that could proceed only if gas existed nearby during their formation process. Astronomers can date this formation to an era when the early sun already shone strongly, preventing giant planets from forming nearby. How, then, could a planet as large as Jupiter, or even larger, orbit its star at a distance less than one-fifth of Mercury's distance from the sun?

At least part of the answer seemed clear: Planets must migrate. If we took Jupiter today and moved it to a distance comparable to 51 Pegasi b's distance from its star ($\frac{1}{100}$ of Jupiter's current distance from the sun), its immense gravitational force would allow the planet to retain most of its gas, even though it would bake at a couple of thousand degrees Celsius. The fact that this first exoplanet to be discovered proved to be a "hot Jupiter" could have an arguably simple explanation: Form a planet at a large distance from its star, and then make it migrate to a much smaller distance.

This scenario left two items crying out for further explanation. What makes planets migrate, and what stops them before the planet plunges into its star? The mechanisms that explain planetary migration and its halt, described in Chapter 11 in some detail, inspire serious debate, but the basic concept of forming "hot Jupiters" far from their stars and having them migrate to much smaller distances has gained general acceptance.

With the number of exoplanets revealed by the radial-velocity technique now rising toward 1,000, we may draw some important statistical conclusions. First, as shown in Figure 8, these planets exhibit an enormous variety in their masses and orbital sizes. Their masses range from a mass comparable to Earth's up to thousands of times Earth's and fifty times Jupiter's. (As discussed in detail in Chapter 6, at the high end of this range, we encounter the boundary that separates an extremely high-mass planet from a "brown dwarf," a failed star whose formation process apparently resembled a star's rather than a planet's.) We should repeat our qualification that all these masses represent minimum values, because we observe only part of a planet's effect on its star. In addition, these results fail to include the lowest-mass exoplanets, because our current radial-velocity techniques cannot reveal planets with masses significantly lower than the Earth's.

Figure 8 emphasizes the fact that although the radial-velocity method has revealed numerous planets orbiting their stars at distances only 10 percent, or even less, of the Earth–sun distance, many more planets move in much larger orbits. The exoplanet orbits cover a wide range of sizes, limited at the large end by two factors. Finding planets with long-period orbits requires long periods of time. The planets moving in such orbits exert comparatively smaller gravitational forces on their stars, which induce smaller changes in radial velocity that can reveal the planets' existence. Exoplanets' currently known orbital periods span nearly 7 orders of magnitude, from 3.5 hours for the planet KOI-1843 b to 2,000 years for

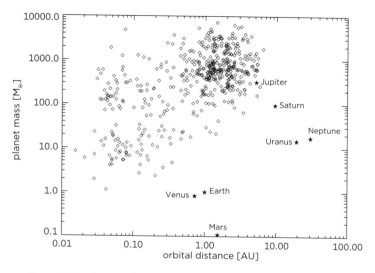

Figure 8 This diagram plots the minimum masses (in terms of the Earth's mass) and orbital distances (in terms of the Earth–sun distance, or AU) of 426 planets discovered by the radial-velocity technique. Note that the axes are logarithmic, expanding by factors of 10. The sun's seven largest planets are also plotted for comparison. (Courtesy of Joshua Winn, from data available at exoplanet.org)

11 Ophiuchi b, whose average distance from its star equals about 250 AU. Many of these exoplanets have almost circular orbits, but many others move in noticeably, or even tremendously elongated trajectories, with the current record holder, HD 20782 b, changing its distance from its star by a factor of about 65 through the course of its 20-month orbit.[5]

Proxima b: The Closest Exoplanet

In August 2016, the radial-velocity technique scored a major public-relations success (and a good one for astronomy as well) with the announcement that astronomers had found an Earthlike planet in orbit around the closest star to the sun. The Alpha Centauri triple-star system, which is 10,000 times more distant than the sun's outermost planet, Neptune, contains two sunlike stars, Alpha Centauri A and B, along with Alpha Centauri C, a faint red dwarf also named Proxima Centauri because it ranks as the sun's closest neighbor. The A and B stars maintain a separation of 24 AU as they orbit their common center of mass every 90 years, so we could model the double-star system with our solar system, if we replace Neptune with a sunlike star. Proxima Centauri, currently at 13,000 AU from its two overlords, follows a highly elongated, half-million-year trajectory around the center of mass of the three-star system. Because it happens to lie on our side of the system's center, Proxima now finds itself at a distance of 4.24 light years from the solar system, slightly less than the 4.37 light years of its gravitational masters. At 1.1 and 0.9 solar masses, respectively, the A and B stars govern the gravitational situation: Little Proxima has only one-eighth of the sun's mass and $1/10,000$ of the sun's luminosity, placing it firmly among the low-mass, low-luminosity stars described in Chapter 9, which dominate the cosmos in numbers if not in brightness. (In astronomical usage, a star's *luminosity* specifies the star's total energy output per second, including all types of radiation such as visible light, infrared, and ultraviolet. The luminosity remains independent of the star's distance from an observer, which affects the star's *brightness,* designated as the star's "apparent brightness" by astronomers.)

As Ralph Waldo Emerson noted, "for every thing you have missed, you have gained something else."[6] Unlike the A and B stars, whose joint output makes them the third-brightest star in the night sky (though they remain forever invisible from mid- or high northern latitudes), Proxima Centauri's low luminosity renders it invisible to the unaided eye, close though it may be. A low stellar mass does make exoplanet detection via the radial-velocity method easier, because the star responds more readily to a planet's gravitational tug. Furthermore, any planets that form along with a low-mass star are likely to have much smaller orbits than those around more sunlike stars, so those planets can therefore affect their low-mass stars' motions more significantly.

During the early years of the twenty-first century, using telescopes in Chile run by the European Southern Observatories (ESO), a consortium of 15 European nations and Brazil, astronomers found hints of variations in Proxima Centauri's radial velocity. In 2016, a team led by the Spanish astronomer Guillem Anglada-Escudé obtained much better radial-velocity measurements with the High Accuracy Radial Velocity Planet Searcher (HARPS), a second-generation spectrographic instrument installed on the 3.6-meter telescope at ESO's La Silla Observatory, which has now revealed more than 150 exoplanets.

HARPS can achieve a precision slightly better than 1 meter per second in its velocity measurements, sufficient to reveal that Proxima Centauri has a planet moving in an 11.2-day orbit that causes the star's radial velocity to vary by 1.38 meters per second from its average value. The amount of these variations implies that the planet has a minimum mass of 1.27 Earth masses (as noted earlier, the mass almost certainly exceeds this value, which would arise only if our line of sight happened to coincide with the planet's orbital plane). Proxima Centauri's 11.2-day orbital period implies a planet–star distance slightly less than ¹⁄₂₀ of an astronomical unit.[7] At this distance, the sun would scorch any planet, but in orbit around its low-luminosity star at 0.05 AU, the planet receives about 65 percent as much stellar energy each second as the Earth does. As a result, the planet's average temperature (if it lacks a heat-trapping atmosphere) should equal 234 K or −39 C. (Scientists often employ the Kelvin, or absolute, temperature scale, denoted by K, whose units have the same size as those of the Celsius scale, but which sets its zero point at absolute zero, or 0 K, equal to −273.15 Celsius, −459.67 Fahrenheit.)

Having shortened Proxima Centauri b's full designation to Proxima b, astronomers revel in their knowledge that the nearest exoplanet, similar to the Earth in mass and presumably in size, orbits just inside the star's "habitable zone" (see Chapter 12). Proxima b confirms the already established conclusion that most stars have planets, and it offers the current first target for futuristic probes that may someday—perhaps even within some of our lifetimes—speed from Earth to send us close-up images of this sizable planet around the little nearby star. We will examine this intriguing possibility in the final chapter of this book.

The Great Recognition: Planets around Red Dwarf Stars

Astronomers found Proxima b several years after fully internalizing the fact that for decades, exoplanet searches had tended to overlook a crucial aspect of the stars that populate the Milky Way. Imbued, perhaps, with the human prejudice to favor our solar system, and aware that before 1995, the sun's family offered the only example of a planetary system surrounding a star, astronomers had directed nearly all their exoplanetary attention toward stars that resemble the sun in their basic properties: their masses, luminosities, and sizes. Eventually, astronomers recognized that the most easily detectable planets do not orbit sunlike stars but instead can be more easily found around the far smaller, far less massive, far less luminous, and far more numerous red dwarf stars. Red dwarfs, known to astronomers as "M stars" from astronomers' strange (yet historically justified) system of stellar classification, are among the least massive of all stars.

Ranking near the extreme low end of the stellar mass range, M stars can barely be distinguished from the failed, would-be stars called brown dwarfs (see Chapter 6). Small, dim, and lightweight though they may be, M stars dominate all other types of stars in their numbers: In our galaxy, for every star with a mass similar to the sun's, 5 to 10 times more M stars dot the starry realm. Astronomers peering through the Milky Way to observe faraway galaxies have been known to refer to these red dwarfs as "vermin of the sky," because the pointlike images of the most distant galaxies can be easily confused with M stars. The fact that M stars have much smaller masses and luminosities than the sun's makes their planets much easier to detect using at least three different techniques: (1) direct observation, because the star does not outshine the planet so severely (Chapter 6); (2) the radial-velocity method, which depends on the amount of a planet's gravitational influence on its star, which grows larger when the star has a smaller mass; and (3) the "transit" method, which reveals planets that pass in front of their stars, since a planet will block the light from a larger fraction of the disk of a smaller star (see Chapter 5).

Once astronomers began to grasp these fundamental truths and to search for planets around red dwarfs, their successes multiplied like fireflies. A recent study estimates that on the average, at least every other M star has an Earthlike planet orbiting within its habitable zone, that is, at a distance from the star likely to give the planet a surface temperature between 0 and 100 C (see Chapter 12).[8]

In the search for exoplanets around M stars, two sets of survey telescopes play a key role. The Harvard-Smithsonian Center for Astrophysics maintains one of them, named MEarth—"M" for red dwarfs, and "Earth" for the attempt to find an Earthlike planet around one of these decidedly non-sunlike stars. MEarth's European counterpart, MASCARA (Multi-site All-Sky CAmeRA), which began full operation in 2017, employs similar arrays of telescopes at the Roque de los Muchachos Observatory on La Palma in the Canaries and at the La Silla Observatory in Chile. Within a few years, these surveys should yield an increasing number of intriguing planetary systems around the coolest stars of the Milky Way.

Further Improvements in Radial-Velocity Measurements

As the twenty-fifth anniversary of the discovery of the first exoplanets revealed by radial-velocity measurements approaches, new and improved spectrographs on ground-based telescopes, capable of finding exoplanets with lower masses and greater planet–star distances than previously possible, will soon be in use. One of these new instruments will begin operation in 2019 at the Keck Observatory, located on the 4,205-meter summit of Mauna Kea, the tallest extinct volcano on the Big Island of Hawaii. Created by Caltech's Andrew Howard along with a host of astronomers and technologists, the Keck Planet Finder (KPF) incorporates a number of improvements, including Zerodur low-expansion glass; better temperature control, in order to keep thermal changes to a minimum; and operation of the spectrographic instruments within a vacuum. Profiting from the immense light-gathering power of one of the Keck Observatories' twin 10-meter mirrors, the KPF will aim to achieve a precision of 30 centimeters per second in measuring stars' radial velocities.[9]

Another new planet-hunting spectroscopic system will soon begin its radial-velocity search for exoplanets from an observatory closely associated with the study of the sun's planets. In 1894, the eminent Bostonian Percival Lowell, who was fascinated by the planet Mars, established the Lowell Observatory in Flagstaff, Arizona. Lowell equipped the observatory at his own expense with an excellent 24-inch (0.61-meter) refracting telescope made by Alvin Clark & Sons, one of the leading manufacturers of fine optical systems at the time. The Clark refractor remains a teaching and public-viewing tool, but the Lowell Observatory's chief astronomical instrument, the Discovery Channel Telescope, now occupies a dark-sky site in the Coconino National Forest, 65 kilometers from Flagstaff near Happy Jack, Arizona. This telescope, funded by a $25 million endowment that secured rights to the instrument's name and the first usage of its astronomical images, and completed in 2015, has a mirror 4.3 meters in diameter, less than half of the mighty Keck's but sufficiently large to achieve—if all goes well—the goals set forth by Debra Fischer from Yale University, which operates the telescope in concert with three other U.S. universities.

Fischer, one of Geoff Marcy's former students, now leads the Yale Exoplanet Laboratory. In collaboration with Colby Jurgenson, she has developed the EXtreme PREcision Spectrometer, or EXPRES, which was installed in Arizona during the late spring of 2017. Yale has the rights to use the Discovery Channel Telescope for 70 nights per year, giving Fischer dreams of the flood of data that should emerge as EXPRES allows measurement of stellar radial velocities to a precision much better than 1 meter per second, perhaps even 10 or 20 centimeters per second.[10] As a reference point, we may note that an Earthlike planet orbiting at 1 AU around a sunlike star induces radial-velocity deviations from the average value of 9 centimeters per second.

Achieving the goals of the Keck Planetary Finder and the EXPRES system will require separating the "noise" in radial-velocity observations described earlier from actual changes in the motions of a star under the influence of one or more planets. Like the Keck Planet Finder, EXPRES will deal with these problems by placing its complex spectrographic system within a vacuum chamber, where its temperature can be controlled to within a thousandth of a degree in order to eliminate any thermal expansion.

The European entry in this array of improved spectroscopic instruments designed to search for exoplanets' effects on their stars' radial velocities carries the name ESPRESSO (Echelle SPectrograph for Rocky Exoplanet and Stable Spectroscopic Observations)—dangerously close to the EXPRES acronym. Fortunately named or not, ESPRESSO, which began operation in 2018, profits from its ability to analyze starlight collected by all four of the 8.2-meter Very Large Telescopes at the Paranal Observatory in Chile, sited on a 2,635-meter-high peak above the Atacama Desert.

If these instruments prove capable of measuring radial velocities to a precision far below 1 meter per second (the speed of a slow walk), and if they can even close in on 10 centimeters per second (slightly faster than a sloth)—as Fischer, Howard, and their collaborators and friendly rivals dearly hope—the floodgates will open for the radial-velocity method to reveal exoplanets with masses similar to the Earth's. Time will tell, however, whether the motions and changes occurring on stars' surfaces impose a definitive barrier, somewhere below 1 meter per second, on astronomers' ability to employ the radial-velocity technique to find Mars-sized and smaller exoplanets.

5

.

FINDING EXOPLANETS
BY THEIR TRANSITS

D uring the first decade of the twenty-first century, as exoplanet hunters used radial-velocity measurements to find
planets with masses as small as the Earth's around a host
of stars, another group of astronomers achieved this same goal by
employing a different technique: the transit method. This method
relies on the fact that the existence of vast numbers of exoplanets
makes it likely that some of them have orbits that carry them directly
across our line of sight. Such a trajectory will cause a temporary dip
in the star's brightness, which will recur each time that the planet
completes an orbit. Astronomers, always lovers of vintage termi-
nology, use the word "transit" to describe the passage of one object
directly in front of another, as seen by a particular observer.

Once astronomers determine that such a brightness decrease
arises not from a starspot, a starquake, or a full-star shudder, but
rather from the transit of an exoplanet, they can measure the planet's
orbital period directly, and thus, thanks to Newton's laws of motion and their estimate of the star's mass, they can determine the
planet's distance from its star. In addition, with a proper estimate

of the star's diameter, astronomers can use the amount by which the star's brightness declines during the planet's transit to find the exoplanet's size, and—based on this size—a reasonable estimate of the planet's mass. (Note that a knowledge of the star's mass and the planet's orbital period still leaves the planet's mass undetermined, because the planet's orbital period does not depend on its mass.)

During the early years of this century, astronomers used the transit method to make several hundred successful planetary detections with ground-based telescopes. They achieved even greater successes employing the transit method following the launch of the CoRoT and Kepler spacecraft, which produce better data from operating in space. Created by the French Space Agency and the European Space Agency, CoRoT, which stands for COnvection, ROtation et Transits planétaires, (or, in English, COnvection, ROtation, and planetary Transits), was launched from the Baikonur Cosmodrome in Kazakhstan and entered an orbit around the Earth at the end of 2006. CoROT proceeded to discover transiting exoplanets until its computers failed in November 2012. During its five years of operation, CoRoT, which could measure brightness variations by 1 part in 10,000, detected about three dozen exoplanets, opening the path to spaceborne planetary discovery.[1]

NASA's Kepler spacecraft, named after the famous seventeenth-century astronomer Johannes Kepler and launched in March 2009, built upon CoRoT's success and embodied a more ambitious plan: to include a telescope with a mirror measuring 1.4 meters in diameter (far larger than CoRoT's 27 centimeters), and to observe the heavens far from Earth's interference. Kepler passed long years in design and preparation as the technological child of William Borucki, an astronomer at the Ames Research Center in California who emphasized to NASA that a great advantage would be gained from placing a telescope outside the atmosphere to search for exoplanets' transits.

All Earthbound observers, submerged within their life-giving, mobile blanket of air, remain forever barred from a completely clear view of any celestial object, no matter how large their telescopes. Invariably, even on the clearest nights, Earth's rippling atmosphere continuously diverts any beam of starlight from following what would otherwise be a perfectly straight path. Instead, different portions of the turbulent atmosphere effectively act as individual lenses, producing what astronomers call "atmospheric refraction" by bending starlight. This bending, or refraction, changes a star's apparent brightness by tiny, ever-varying amounts on timescales as short as milliseconds. Observed on timescales shorter than one second, these changes produce the "twinkling" of starlight so beloved by poets but disrespected by astronomers. The first of these effects prevents Earthbound astronomers from determining a star's position with an accuracy much better than about one second of arc, while the second introduces an uncertainty in the star's brightness by a modest fraction of a percent, which naturally frustrates astronomers who seek to measure changes in brightness by the same fraction, or even less.

Despite these handicaps, devoted astronomers had opened the door to exoplanet discoveries with the transit method by finding transiting planets with Earthbound telescopes. In 1999, astronomers announced the first exoplanet discovered by the transit method, in orbit around the star HD 209458. Described in detail in Chapter 9, this success, achieved by two groups of astronomers, represented the fruit of a decades-long search for exoplanets, led primarily by the Harvard astronomer David Latham.[2] In space, the difficulties imposed by the Earth's atmosphere evaporate. All that remains (to simplify greatly) is to search for situations in which a planet's orbit happens to align with our line of sight.

From the view of the solar system available on the sun's third planet, only two objects undergo noteworthy transits: Mercury, which passes directly in front of the sun every few years, and Venus, which does so about twice in every century, thus placing its transits among the rarer astronomical events. Observations of Venus's transits from different points on Earth once offered the best means of determining the scale of distances in the solar system. In 1761, as the French astronomer Guillame Le Gentil sailed toward India in order to observe a transit of Venus, his ship encountered such severe weather that he found himself still at sea during the event, his instruments rendered useless by the motion of the waves. Instead of taking the long journey back to France, he scouted for the best location to observe the transit of 1769 and settled on his original site at Pondicherry, where, after a run of fine weather, clouds covered the sun on the crucial morning. Depressed, Le Gentil gathered the strength to return home, where he learned that he had been declared dead and that his wife had remarried. Legal action and the king's intervention eventually restored his original official position, but his story remains a favorite among astronomers who recognize the many unexpected obstacles to achieving observational success. This century's Venus transits occurred in 2004 and 2012, leaving future astronomers in eager expectation of the next two, in 2117 and 2125.

More fortunate than Le Gentil, the modern-day astronomer Bill Borucki spent years promoting a plan to build on CoRoT's success by creating a spacecraft with a much larger telescope that would be sent far from Earth and devoted to observing the transits of exoplanets. If we use the inner solar system as a model, Borucki's vision centered on using a spacecraft that could find a Venus or an Earth (or a larger planet) by observing the decrease in sunlight that the planets' transits across the face of their star would cause. (We'll leave Mercury out of the discussion because of its much smaller size.) Venus and Earth have diameters about $1/100$ of the sun's, and therefore cover an area on the sky about $1/10{,}000$ of the sun's, so their transits would cause a dip in the sun's light by 1 part in 10,000—which would be unobservable from Earth's surface, but a surmountable problem in space. Kepler aimed to detect bright-ness changes by 20 parts per million, about 50 times better than the smaller CoRoT satellite could. As things turned out, the spacecraft's actual performance reached 29 parts per million, still an impressive advance in the search for transiting planets.

The basic problem with observing transits from space—once we ignore the issues arising from the authorization, design, funding, construction, launch, and operation of a distant spacecraft, along with the complex data reduction involved in the continual moni-toring of a multitude of stars in hopes of detecting planetary transits and confirming that they do not arise from starspots, starquakes, or other changes in the star's brightness—resides in the rarity of these events. To wait a century for any astronomical event to occur would test all human patience, not to mention that of the mem-bers of Congress whose assent would be required to secure the necessary funding.

The solution was obvious: Observe not one star but thousands of them. If a spacecraft could study 36,500 stars, the intervals be-tween planetary transits across the face of one of those stars would decrease by the same factor, reducing the time between transit observations from one century to one day. The basic problem re-mains: Only a small fraction of all planetary systems will happen to have orbits that carry them across our line of sight to their stars—the only possibility that allows us to see a transit. But num-bers don't lie: By observing a large number of stars simultaneously, we greatly increase the rate at which we can detect planets by the transit method.

The first decade of the new millennium saw NASA contractors and engineers create the Kepler spacecraft under the comparatively low-budget "Discovery Program," which included spacecraft missions to investigate Mercury, Mars, three different asteroid re-gions, two comets, the moon, the solar wind, interstellar dust, and—by far the most outgoing in its observational reach—the

Kepler spacecraft, which as David Charbonneau put it, "blew the transit doors wide open."[3]

Designed for the single purpose of detecting the transits of exoplanets and constructed at a cost of $600 million, Kepler carried into space a 480-kilogram payload that included a 1.4-meter reflecting telescope, along with the optical and photometric equipment needed to allow the telescope to monitor stellar brightnesses with high precision. Astronomers chose to send the Kepler spacecraft to a distance from the sun a few percent greater than ours, where its Earth-trailing orbit has caused it to increase its distance from us by about 25 million kilometers per year.[4]

During its first four years of operation, Kepler's wide-angle telescopic system, working in the dark of space, freed from the Earth's atmosphere and far from any Earthglow, inspected the star-rich region of the sky where the constellations Cygnus, Lyra, and Draco meet. Its target area spanned about 115 square degrees—"two scoops of the Big Dipper," as its masters liked to say—a region more than 60,000 times larger than the Hubble Space Telescope's field of view. Within this area, and unhindered by our cycle of day and night, the Kepler spacecraft recorded the light from 156,000 stars with a "cadence" of 30 minutes. In other words, amazing though it sounds, Kepler measured the brightness of each of these stars at half-hour intervals. Data from the photometric measurements, sent to Earth via radio, received prompt analysis that searched for periodic dips in star brightness that could signal the existence of one or more planets, moving in orbits that carry them across our lines of sight to their stars.

Overcoming Stellar Variation

The brief diminutions in starlight during a planetary transit typically last for a few hours, the time it takes for the transit of Venus across our sun. In attempting to find exoplanets by their transits, astronomers knew that they confronted a serious problem. No star maintains a perfectly steady energy output, though they do so almost perfectly for millions of years. Because the rate of nuclear fusion in their cores depends strongly on the core temperatures, stars possess a self-correcting mechanism that keeps their energy output nearly constant. If a star's core grows a bit hotter, its rate of nuclear fusion increases, releasing extra energy. This pushes the core to a slightly greater size, lowering its temperature and bringing the nuclear-fusion rate back to its initial value. Conversely, if the core grows somewhat cooler, it has a slightly lower ability to maintain its size against the force of the star's overlying layers. The pressure from these layers shrinks the core slightly, increasing its temperature and thus the rate of nuclear fusion.

The energy released in a star's core makes its way toward the surface through an enormous number of collisions among the particles that constitute the star's innards. Near the surface, where the energy has a chance of escaping into space in the form of photons that form visible light and ultraviolet or infrared radiation, irregularities have a better chance of developing. These irregularities sometimes include the stellar flares that temporarily and significantly increase the energy radiated into space, without any direct connection to the nuclear-fusing region at the star's center. At the star's surface, cooler areas can develop and persist for a few days or weeks, producing "starspots" analogous to the "sunspots" on our own star's surface.

These stellar flares, starspots, and other effects can make any star's light decrease or increase—sometimes for a few hours, sometimes for longer periods. But astronomers eager to employ the transit method have a nearly certain means of distinguishing any such effects from the transits of one or more planets in orbit—one that arises from the clockwork nature of orbital dynamics. A single diminution in starlight captures astronomers' awareness. A second one that occurs days, weeks, or months later calls for additional attention. And a third, spaced by the same interval in time as that between the first two, provides a nearly definitive signal of an object moving in orbit, an object whose existence can then be verified with successive diminutions in starlight.

The stars that Kepler studied lie in the same region of the Milky Way as the sun, and therefore have distances from the center similar to our own sun's distance from that center, some 26,000 light years. The stars' distances from the solar system range from about 600 to 3,000 light years, with the far boundary imposed by the diminishing brightness of the light from progressively more distant stars. Most of these stars have intrinsic luminosities (energy output per second) similar to the sun's, because stars with much greater luminosities are comparatively rare, and stars with much lower luminosities were too faint for Kepler to record.

Four Years of Kepler's Exoplanet Discoveries

During the four years of the Kepler mission's original planned lifetime, the spacecraft found more than 4,000 planet candidates around a small minority of the 156,000 stars under continual scrutiny. Closer inspection confirmed more than half of these (2,335) as actual planets. Their orbital periods range from 8.5 hours up to 3.6 years, and their orbital sizes cover a correspondingly wide range, placing the exoplanets at distances of 0.01 to 2.7 AU.

How should we extrapolate from these results to estimate the total number of planets around the stars that Kepler observed? Broadly speaking, if planets' orbital orientations are randomly distributed, the chance that we will observe a transit of an Earth-sized planet in an Earth-sized orbit around a star similar to the sun is about 1 in 200. Thus Kepler's 2,343 planets imply that roughly half a million planets exist around the stars that the spacecraft studied. (Some complexity in this extrapolation arises from multi-planet systems: If one planet undergoes a transit, they probably all will.) If we keep multiplanet systems firmly in mind, we can summarize the most basic result from Kepler's first four years as follows: *the majority of sunlike stars have one or more planets in orbit around them.*[5]

Sizes and Masses of the Kepler Planets

When we derive the sizes of planets by observing the percentage of the light from their stars that the planets block as they transit, some uncertainty must arise from our imprecise knowledge of the sizes of their stars themselves. Working from a century of observational and theoretical determinations of stars' basic characteristics, astronomers can make a reasonable estimate of a star's diameter from observations of the star's spectrum. Kepler's success emphasized the need for more precise estimates of the stars observed to have one or more transiting planets. Using the largest Earthbound telescopes, astronomers have improved the accuracy of their knowledge of stellar diameters, and therefore of the diameters of the planets that perform transits across them.

To a first approximation, a planet's size provides a reasonably good estimate of its mass, because planets about the size of the Earth should be made mostly of rocks and metal, with a density three to six times the density of water, and much larger planets should resemble Jupiter and Saturn, with densities comparable to water's. By adopting this approach, astronomers could create the array of planetary masses and orbital sizes shown in Figure 9, which updates Figure 8 by adding the planets that Kepler found, along with a relatively few planets disclosed by direct observation (Chapter 6) and by their gravitational lensing of starlight (Chapter 7).

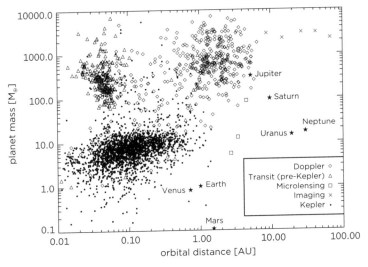

Figure 9 This updated version of Figure 8 adds 2,335 Kepler planets discovered by the transit method, together with 16 planets found by the microlensing technique, and a few planets, all with masses much greater than Jupiter's 318 Earth masses, orbiting far from their stars, seen with direct imaging. The transiting planets discovered before Kepler's operation are so large that they could be detected from Earth. (Courtesy of Joshua Winn, from data available at exoplanet.org)

Some transiting exoplanets offer astronomers the happy concurrence of producing observable and measurable variations in radial velocity. As we have seen, in most radial-velocity-detection situations, we can deduce only the planet's mass divided by a number less than or equal to 1, which depends on the orientation of the planet's orbit with respect to our line of sight. For a transiting planet, however, that factor equals 1: We then observe the planet's entire gravitational effect on its star.

With this advantage, astronomers can find both the size of the planet (from the amount by which the transit reduces the star's light) and its mass (from the radial-velocity changes of the star's motion). Astronomers first applied this powerful combination of the two chief methods for detecting and studying exoplanets in their discovery of the planet HD 209458 b, also known as Osiris (see Chapter 9). They have already achieved a significant number of successes in this effort, and astronomers look forward to the next few years, when new spacecraft and extremely large telescopes will increase this number impressively.

The discovery of transiting exoplanets, along with the ability to estimate the planets' masses from their sizes (in turn derived from the diminution of starlight observed during their transits), allowed astronomers to put to rest an earlier hypothesis: namely, that almost all the exoplanets discovered from radial-velocity observations were high-mass objects that happened to move in orbits almost perpendicular to our line of sight. Such orbits tend to indicate a small minimum mass for the exoplanet. Now, however, astronomers have many definitive examples of comparatively low-mass planets whose orbital planes coincide with our line of sight.

Studying Exoplanet Atmospheres with Transit Observations

Planetary transits provide information that extends well beyond the basic facts of an exoplanet's size and orbit. With sufficiently powerful instrumentation, astronomers can compare the details of a star's spectrum observed during a planetary transit with the spectrum measured at other times. If the planet has a detectable atmosphere, astronomers will find differences that the planet's atmosphere produces. In most cases, the planet's atmosphere absorbs the star's light at certain wavelengths—wavelengths that signal the presence, and the abundances as well, of common atoms and molecules such as sodium, water, carbon monoxide, and ammonia. The relative abundances of these molecules can indicate the temperature within the portion of the planet's atmosphere re-

sponsible for most of the absorption of starlight. In this way, planetary transits' effects on starlight can reveal the atmospheric compositions and temperatures of exoplanets many light years away
that would otherwise remain unglimpsed.[6]

As Kepler Loses Its Wheels,
K2 Emerges from Failure

The Kepler spacecraft's telescope focuses light onto a single instrument, a photometer that measures the brightnesses of stars with
high precision. To direct its telescope toward the myriad stars in
its program, Kepler relied on four comparatively heavy, rapidly
rotating "reaction wheels"—essentially large gyroscopes—that allowed the spacecraft to maintain its orientation and to move
accurately from one field of view to another. Any spacecraft similar to Kepler requires three reaction wheels, one for each direction in space, to achieve this goal. The engineers who built the
Kepler spacecraft estimated its useful lifetime at four years; this
was largely based on the reaction wheels' expected duration. Kepler
began its mission in March 2009 with four functioning wheels. In
2012, one of them failed; then, in May 2013, just past the four-
year mark, the failure of a second reaction wheel left the spacecraft unable to continue its observations.

Or did it? The engineers at NASA's Ames Research Center,
working with those at the Ball Aerospace Division who built the
spacecraft, hit on an ingenious maneuver that would allow the
spacecraft to continue its search for exoplanets in a highly original
manner. Instead of using a third reaction wheel, the Kepler spacecraft now uses the sun to maintain its proper orientation. Kepler
senses the weak pressure from solar radiation, which tends to push
the spacecraft away from a particular line of sight. The engineers
realized that by orienting the spacecraft to distribute this radiation

pressure evenly across Kepler's surface, they could maintain the spacecraft in a particular orientation with respect to the sun, and then use the two remaining reaction wheels to search in the two additional directions. This orientation requires that the Kepler spacecraft must observe the sky only in directions close to the ecliptic, the apparent path of the sun around the sky throughout the course of a year. This path lies within the 12 constellations of the zodiac, which include a wealth of stars to search for possible planets.

Thus the Kepler mission achieved a second life, under the new name K2. The new mission began in May 2014, and it should last for at least four years, as the original mission, now called Kepler Prime, did. The NASA scientists divide the zodiacal region around the sky into separate fields of view, each as large as the one that the original mission studied for nearly four years, and they perform shorter "campaigns" about 80 days long to observe each of them, thus covering four or five fields of view per year.[7]

The K2 program not only examines regions of the sky that are far from those studied by the original mission, but it also concentrates on a different type of star in its hunt for exoplanets. While Kepler Prime's focus on sunlike stars formed the bulk of the original mission's subjects, just 40 percent of K2's targets qualify as M stars, the red dwarfs that are now recognized as prime planet-hunting territory because (a) they have plenty of planets and (b) their small sizes make it easier to detect small planets around them. The low luminosities of M stars require Kepler to observe them at comparatively nearby distances. Fortunately, this does not pose a serious problem, thanks to the ubiquity with which M stars have scattered themselves throughout the Milky Way. Despite their low luminosities, the K2 stars lie so much closer to us than those observed by Kepler Prime that they typically have 5 to 10 times the brightness of the stars in the Kepler Prime field. This makes ground-based spectroscopic studies of the K2 stars an easier task than observations of the spectra of the light emitted by the Kepler Prime stars.

Since 2014, K2 has found a few hundred exoplanets around nearby M stars, and it has identified an equal number of candidates now waiting to be verified through further analysis of the flood of data from the revived Kepler mission. The orbital periods and distances of these exoplanets exhibit a wide range, from 4.3 hours and 0.006 AU up to 45 days and 0.21 AU, with planets more distant than this from their stars still lurking in the data remaining to be analyzed. Astronomers can already state, as they extrapolate from transiting exoplanets to the total planet population, that almost every cool star has at least one planet with an orbital period shorter than 50 days (one might reasonably expect exoplanets around small, low-mass stars to move in such comparatively small orbits), and a mass between 0.5 and 4.0 times the Earth's.[8]

In addition to searching for exoplanets around M stars, the K2 mission has found exoplanets around stars less than 100 million years old—that is, ages less than $1/45$ of the sun's—in nearby star clusters. Some of these exoplanets qualify as "young, hot Jupiters" that take only a few days to orbit around their stellar hosts. K2 has also studied star-forming regions, galaxies beyond the Milky Way, and numerous small objects within the solar system, including asteroids with Jupiter-like orbits and some of the trans-Neptunian objects whose size and numbers have, at least for now, downgraded Pluto from planetary status. If all goes well, K2 should at least double the number of Kepler planets.

6

·

DIRECTLY OBSERVING
EXOPLANETS

Astronomers have employed four fundamentally different approaches in their successful searches for exoplanets: radial-velocity measurements, transits, gravitational lensing, and direct observation. Among these four, direct imaging (to use its astronomical name) ranks far below the first two in achieving success, for the excellent reason that any exoplanet's weak reflected light tends to be lost in the glare from its nearby star. On the one hand, we may marvel that any exoplanets have been found by direct imaging. On the other, we anticipate that the giant new ground-based telescopes described in Chapter 13 will soon change this balance and elevate direct imaging into the primary ranks of successful approaches to making exoplanet observations. Eventually, the WFIRST spacecraft, also described in Chapter 13, may provide even better direct-imaging results.

The enormous brightness difference between a star and its planet in visible light leads astronomers to observe in other wavelength domains their attempts to obtain direct images of exoplanets. In particular, astronomers incline toward the infrared portion of the spectrum, where planets emit comparatively large amounts of radiation. Any object with a temperature of a few hundred degrees above absolute zero will emit most of its energy in the form of infrared radiation. With infrared eyes, we would see each other and all objects on Earth glowing in infrared, rather than relying on visible light reflected from those objects to locate them. In astronomical circles, our planet likewise glows in infrared, a sign of temperatures comfortable for life (see Chapter 12). But the Earth's atmosphere absorbs infrared radiation over a wide range of frequencies and wavelengths, making infrared study of the cosmos a difficult undertaking from ground-based observatories. Some infrared wavelengths and frequencies can penetrate at least part of the atmosphere and reach giant telescopes sited at high altitudes, of which the Mauna Kea Observatory in Hawaii stands tallest at 4,205 meters (13,796 feet).

Direct Imaging of the Largest Exoplanets

Throughout the long years of fruitless searches, and even during the first decade of exoplanet discoveries, astronomers' hopes of directly observing exoplanets remained a dream that might not become reality in their lifetimes. Everyone knew the problem: In visible light, a star outshines its planets by roughly a billion times, and even in the more favorable domain of infrared, in which stars are typically dimmer and planets shine far more brightly, the star-to-planet brightness ratio drops only to about one million, or, in extreme cases, to a mere 100,000. In addition, a much greater obstacle to observing planets directly arises from the astronomically tiny distances between stars and their planets.[1]

Despite these obstacles, astronomers have managed to obtain images of nearly 100 exoplanets. Before we examine these successes, we should stress that in this context, the word "image" does not refer to a map or a photograph with details (no matter how imperfectly visible) of the planet's surface or atmosphere, but rather to a single dot of light, happily distinguishable from the much brighter dot nearby. This low-information content remains highly welcome: Even a single-dot image of an exoplanet allows the possibility of direct spectroscopic analysis of the planet's infrared radiation. Because spectroscopy remains astronomers' most important tool for analyzing the radiation from any cosmic object, it plays a prominent role among astronomers' tools for understanding the nature of exoplanets.

For now, and for the immediately foreseeable future, the exoplanets most suitable for direct imaging are giant planets—the larger the better—that orbit far from their stars. Unlike the transit and radial-velocity detection methods, which favor close-in planets, direct imaging works best for seriously large planet–star separations. The infrared radiation from these planets typically arises from the slow contraction of the planets' interiors, as occurs for Jupiter or Saturn in the sun's planetary system. Younger planets generate more infrared radiation from this effect, making them more ideal targets in direct-observation searches.

Because these attempts involve their own infrared radiation, planets farther from their stars do not suffer a decrease in their brightness, as would occur with planets that shine by the light that they reflect from their stars. In fact, greater planet–star distances make the planets easier to view separately from their stars. These considerations favor finding exoplanets comparatively close to the solar system twice over: The exoplanet's infrared radiation appears brighter to us, and the geometry of the orbital situation creates greater separations on the sky between the planet and its star. The most favored candidates for direct imaging therefore appear in systems that contain young planets with 3 to 80 times Jupiter's mass that orbit their stars at distances much greater than those with which our own giant planets orbit the sun. The top end of this mass range brings us into the realm of brown dwarfs, which are more correctly described as failed stars than as giant planets.

Giant Planets or Brown Dwarfs?

At the high end of the planetary mass scale, astronomers must distinguish between true planets and "brown dwarfs," objects with masses larger than planets but with too little mass to qualify as stars.[2] All stars shine from nuclear fusion in their cores, where they fuse hydrogen nuclei (protons) into helium nuclei. This process transforms some of the energy of mass of the fusing particles into kinetic energy, which slowly works its way to the stars' surfaces, heating the gases there to the point that they emit visible light and other forms of radiation. Nuclear fusion proceeds only when the temperature rises so high that protons move with speeds that allow them to collide so violently that they overcome their mutual repulsion. The required temperatures, which measure the average speed of a large group of particles, come close to 10 million K. True stars generate these temperatures by squeezing their interiors as the result of the mutual gravitational forces among all their particles. Greater squeezing results in higher temperatures, and the least massive and least luminous stars have masses barely sufficient to reach the 10-million-K threshold at their centers.

Astronomers contrast the lowest-mass stars, which do produce energy by nuclear fusion, from their less distinguished cousins, the brown dwarfs, which have masses so small that they cannot perform this most basic stellar function. Instead, the brown dwarfs, forever unworthy of the proud title of star, generate modest amounts of energy through their continuing slow contraction, which heats their interiors (and would, if they only had more mass, eventually turn them into stars). The mass division that separates true stars from brown dwarfs lies at 75–80 times Jupiter's mass, equal to about 7.5 percent of the sun's mass.[3]

Current estimates place the number of brown dwarfs in the Milky Way close to the number of stars, with several hundred billion of each species.[4] The results from recent planet hunting suggest that our galaxy contains at least as many planets as stars, and perhaps several times more planets. (Recall that these numbers leave stars, and presumably brown dwarfs as well, separated by at least a light year from their closest neighbors unless they belong to double- or multiple-star systems.) In view of these enormous numbers and the fact that both large planets and brown dwarfs generate impressive amounts of infrared radiation, what lines of demarcation distinguish brown dwarfs from extremely massive planets? The answer lies, for better or worse, in astronomers' theories of how stars and planets form. As we discuss in Chapter 11, stars form (so accepted models state) through the collapse of clouds of gas and dust at the center of a rotating "protostellar disk" that gives birth to the star and its planetary system. This "top-down" model likewise applies to brown dwarfs, which lack only the mass that would make them true stars. Giant planets, in contrast to stars and brown dwarfs, build themselves in "bottom-up" fashion, gathering gaseous layers around their solid cores, which formed, as rocky planets do, by the accretion of much smaller particles that collided and stuck together.[5]

Two difficulties appear here. First, competing theories assert that the largest planets, and those most distant from their stars, form from the collapse of unstable gas clouds, much like the process that forms stars. These types of planets correspond to those most accessible to direct imaging. Second, even if all planets form from smaller particles rather than in a top-down manner, we have no good way to observe this distinction. Nevertheless, majority opinion places brown dwarfs within the mass range from 13 to

75–80 times Jupiter's mass; in contrast, planets have less than 13 Jupiter masses.

Coronagraphic Masks and Adaptive Optics

Without immersing ourselves too deeply in the details of the brown dwarf–giant planet distinction, let us see what astronomers can find among extremely large planets that orbit far from their stars. Even these planets will not reveal their existence without the application of the significant technological cleverness required first to block the star's much greater emission and then to enable a telescope to counteract the blurring effects of our atmosphere.

The technique of blocking the light from a star without covering a nearby planet carries the now-historic name of "coronagraphy" because its first, and for many years its only, astronomical success lay in blocking the light from the sun's disk to allow observation of the much fainter, thinner, gauzy "corona" that extends for many solar diameters outward from the sun.[6] This corona emits only about one one-millionth as much light as the solar disk does—a ratio similar to the planet-to-star ratio that astronomers must overcome in detecting infrared radiation. The chief differences and problems in securing planetary images by coronagraphy reside in the fact that these stars and planets are many million times farther away than the sun.

Successful solar coronagraphy required years of effort. It was used by telescopes sited at high altitudes, where the thinner atmosphere might remain sufficiently calm to prevent a sliver of the sun's disk from jumping out from behind the "coronagraphic mask" that was placed within the focal plane of a telescope to cover the solar disk precisely. The same analysis applies in spades to the much more difficult coronagraphic observations of exoplanets, which involve coronagraphic masks with complex geometric forms designed to block as much of the star's light as possible while allowing the planet's radiation to reach a telescope's detectors.

Astronomers have risen to these challenges, and they hope soon to improve tremendously on their modest successes once they have an advanced spaceborne coronagraph. The Hubble Space Telescope also incorporates a coronagraphic system, but it functions well below the optimum possible for a spaceborne telescope of its size. Nevertheless, the system points toward a glorious coronagraphic future for telescopes in space (see Chapter 13).

Coronagraphy has currently had limited success. The great new improvement in ground-based telescopic observations of the heavens during the past two decades carries the name of "adaptive optics." This phrase describes optical systems designed to compensate for the ever-changing image distortion caused by the ever-variable atmosphere, which continuously refracts any beam of radiation by tiny amounts. Adaptive optical systems respond by continually measuring the rippling of the atmosphere and adjusting telescope mirrors, on timescales measured in milliseconds, to compensate for the changes in the image that the ripples would induce. This approach requires an optical system that can respond and compensate quickly and appropriately. An adaptive-optics system monitors atmospheric refraction by observing either a guide star in the field of view, or, more often, an artificial guide star generated by shining a laser beam upward and observing either its reflection from atmospheric layers some 20 kilometers high, or the light that the laser induces from sodium atoms at even greater heights. The corrections to the optical system arise in a deformable secondary mirror, onto which the main telescope mirror reflects incoming radiation. With a computer that can receive information from the guide star and direct the machinery that governs the deformable mirror, adaptive optical systems have markedly increased astronomers' ability to push their great telescopes almost to their natural limits—that is, to the capabilities they would have without the distortions introduced by our life-giving atmosphere.[7]

The First Direct Exoplanet Observations

Coronagraphy and adaptive optics now underlie most present attempts (and even future ones), to *see* exoplanets and to measure their apparent brightnesses, and also to study their spectra, from which astronomers can determine the temperatures and compositions of the planets' surfaces or (if they exist) their atmospheres. With their technological advances, astronomers have directly imaged exoplanets that orbit stars with distances of 25 to 500 light years from the solar system. Beta Pictoris, one of the closest stars with a directly imaged planet, first gained fame in astronomical circles when astronomers found that a disk of debris, presumably left over from the formation of planets, surrounds the star. Even today, planet formation may continue around Beta Pictoris, 63 light years from Earth: Astronomers estimate the star's age as only about 20 million years, less than half a percent of the age of the sun and its planets. Next in discovery came the star's directly imaged exoplanet, Beta Pictoris b, which currently qualifies as the directly imaged exoplanet with the smallest orbit around its star.[8]

At a distance of 9 AU from its star, about equal to Saturn's distance from the sun, Beta Pictoris b takes about 22 years for each orbit, but it has an estimated mass that is about 7 times Jupiter's and 24 times Saturn's.[9] The planet's infrared radiation implies a temperature close to 1,600 K, which must arise from its sources of internal heat rather than from incoming radiation from a star 9 AU away. Spectroscopic studies have shown that Beta Pictoris b has an atmosphere, and that the planet rotates once every 8.1 hours, a rotation even more rapid than (much less massive) Jupiter, which has the shortest rotation period (9.9 hours) of the sun's eight planets.

In the final years of the past decade, astronomers employed an advanced adaptive-optics system on two telescopes at the Mauna Kea Observatory, one of the 10-meter Keck twins and the 8.1-meter Gemini North instrument, to study infrared radiation from the vicinity of the star HR 8799. This star, 129 light years from Earth, half again as massive as the sun and five times more luminous, has an estimated age of only 30 million years. The astronomers found four giant planets around HR 8799, each with approximately five to seven times Jupiter's mass, at distances that range from 15 to 68 AU—reminiscent of Saturn's orbit around the sun at a distance of 9.5 AU, Uranus's at 19.2 AU, and Neptune's at 30 AU. They also found two disks of gas and dust around the star; one disk lies inside all four planets' orbits, while the other disk lies outside all four. During the decade since this discovery, continuing observations have recorded the orbital motions of the four planets. As expected from the rules of orbital dynamics, the closer planets move more rapidly than the outer ones, with their complete orbits taking from 45 to about 460 years.[10]

Building on this success, astronomers directed further attention toward HR 8799 and its planets with a specialized instrument at the 8.1-meter Gemini South telescope, situated 2,722 meters high on Cerro Pachón in Chile. More than a dozen institutions in the United States and Canada created the Gemini Planet Imager, or GPI, to work with the Gemini South Telescope. The GPI employs a coronagraphic mask, adaptive optics, and an advanced spectroscopic system that have been used in combination to find evidence for atmospheres around HR 8799 c and d (the second and third most distant of the star's planets), which include molecules of water, carbon monoxide, and methane.[11]

Close to Cerro Pachón, astronomers have deployed an instrument similar to the GPI, which is likewise aided by adaptive optics and a coronagraphic mask, that carries the name SPHERE, an acronym for Spectro-Polarimetric High-contrast Exoplanet REsearch. SPHERE analyzes the light received by one of the four 8.2-meter Very Large Telescopes at ESO's Paranal Observatory. In mid-2017, SPHERE found its first exoplanet, nine times more massive than Jupiter, orbiting at almost 92 AU from the star HIP 65426, 385 light years away.[12]

Because distinguishing extremely large exoplanets from brown dwarfs poses an ongoing problem, some uncertainty exists about the precise number of exoplanets that astronomers have imaged directly. At this writing, the Extrasolar Planets Encyclopedia database (exoplanet.eu) lists 93 of them, but a few of them are peculiar cases that fall outside the traditional definition of direct imaging. The planets' masses range from 3 to 79 times Jupiter's, so the most massive ones lie at the boundary with brown dwarfs. The great majority of these planets have distances from their stars larger than Earth's distance from the sun, with some of these distances ranging to many thousand AU and even beyond. These directly imaged planets, modest though their numbers may be, already tend to confirm the hypothesis that many young stars possess equally young planets much more massive than Jupiter that orbit at distances significantly greater—in many cases far greater—than Jupiter's distance from the sun.

7

•

DETECTING PLANETS
WITH EINSTEIN'S LENS

Perhaps the sweetest, the cleverest, and in some ways the most frustrating technique for finding exoplanets arises from the theory of general relativity, the leap forward in our understanding of the physical universe that made Albert Einstein famous in 1919. This method, which we may call "Einstein's lens," allows astronomers to perceive the existence of exoplanets not by observing the planets' own light, or their stars' light, but instead by detecting the effect that their gravitational forces have *upon* light from far more distant stars.

General relativity theory, which scientists call general relativity, describes the effect of gravity as a bending of space. This concept came as a shock (and still does) to those who feel strongly that space has no ability to bend, and no business doing so. Instead, our intuition insists that space just "sits there," empty and unchanging, no matter what objects or events may come to pass within it. In opposition to this intuitive feeling, which remains strong no matter how scientifically valid his theory has proven to be, Einstein posed a subtler, and, as it turned out, more useful and more accurate description: Space can and does bend under the influence of any object with nonzero mass. Space bends the most in regions closest to any particular object, and it bends more in the presence of more massive objects.

As a prime example of this bending, Einstein predicted that when we observe the light from a distant star that happens to lie almost directly behind the sun, we will find that the sun's gravity makes the rays of starlight deviate from the paths that they would follow were the sun not there.[1] To verify Einstein's prediction, we must measure a star's position on the sky in two different situations, once with the sun present and once with it absent. The latter situation poses no problem, but the former encounters a difficulty analogous to attempts to observe exoplanets directly: the drowning of faint starlight in the sun's nearby glare.

The sole remedy for this problem available in Einstein's time arose in the few minutes of a total solar eclipse, when the moon blocks the light from the sun's disk. This creates a few minutes of near darkness, lit only by the faint glow from the gaseous corona that surrounds the sun. This brief dark time gives astronomers the chance to record the positions of stars that appear close to the sun on the sky and to compare these images with photographs taken at a different time of the year, when the Earth's motion in orbit has caused the sun to occupy an entirely different location on the sky.

In May 1919, a rightly renowned total solar eclipse allowed astronomers to make these crucial measurements, which brought Einstein worldwide prominence by demonstrating that his reinterpretation of gravity had merit. Throughout the next century, a host of similar observations, along with other experiments based on general relativity, have continued to verify, with increasing accuracy, the validity of Einstein's brainchild. From time to time, physicists have proposed modifications to Einstein's theory. Although these suggestions seemed entirely reasonable from a mathematical standpoint, they failed to agree with measured reality.

Sixty-plus years after his death, Einstein continues to stand as the genius who provided our best understanding of the most obvious and mysterious of the forces of nature.

As a wrinkle on the gravity-bends-space principle of general relativity, Einstein noted that in certain astronomical situations gravity could effectively act as a lens, concentrating and magnifying the light from a distant object. Consider what happens if a massive object happens to move across astronomers' field of view, passing in front of a star under observation. (Since all stars in the Milky Way are in constant motion, this happens more often than one might imagine.) If the object passes exactly in front of the more distant one, at the moment of perfect alignment the object's gravitational force will bend light from the distant star all around it, producing what astronomers now call an "Einstein ring." If, instead, the object passes almost, but not quite, directly in front of the star, its gravitational bending of light will tend to create two distorted images of the distant star. And if the line-up differs even more significantly from perfection, astronomers will detect only a single image. In any of these situations, the object's gravitational deflection of the rays from the distant star briefly focuses and concentrates its starlight, making it appear significantly brighter than it does when the object is nowhere in the picture. Astronomers call this effect "gravitational lensing"; the term includes the focusing effect but refers more generally to the increase in a distant object's apparent brightness that occurs when a closer object's gravity bends the light from the more distant one.

Before we consider how gravitational lensing allows astronomers to find exoplanets, we may note that in recent years, astronomers have extended their gravitational-lens observations to the far reaches of the visible universe. They have used the mass concentrated within clusters of galaxies hundreds of millions of light years from Earth as a set of gravitational lenses that sharply increase the brightness of still more distant galaxies and create images with far more detail than would appear without the lensing effect. Einstein would justly be proud.

Within our Milky Way galaxy, gravitational lensing works—in fact, it becomes more easily observable—even if the closer object emits no light of its own. The lensing enhances the light from the more distant star without changing the amount of light received from the closer object. The effects of lensing depend only on the mass of the closer object and how directly it moves across the beam of light from a distant star. Gravitational lensing can be produced by burned-out stars, neutron stars, black holes, or anything else with mass. Because greater masses will induce larger effects, gravitational lensing offers a way to measure the mass of an object, whether seen or unseen, that happens to pass almost directly across the line of sight to a distant star.

Suppose that the closer object happens to be a star with a planet in orbit around it. In that case, astronomers measuring the light from a distant star may see a sizable gravitational-lens effect created by the nearer star, but superimposed on this brightness increase they will observe a much smaller, similar effect that arises from the planet's gravitational force. This secondary effect will depend both on the planet's mass and on the orientation of its orbit, which may cause the planet to pass slightly closer, or slightly farther away, from the trajectory that its star takes across our line of sight to a distant star. In almost all cases, however, the comparison of the sizes of the lensing effects from the star and its planet provides a direct indication of the ratio of their masses.

When Einstein first discussed this effect, he knew that the enormous distances between stars in the Milky Way made it unlikely that astronomers could record the events just described, even for a star. Indeed, astronomers observed no such gravitational lensing in Einstein's lifetime. Massive improvements in their observational capabilities introduced in the past half century, however, have allowed astronomers not only to record gravitational-lensing events, but also to use general relativity to find planets otherwise un-detectable. This marvelous path to planet detection, which astron-omers call "gravitational microlensing," or simply "microlensing," as a reminder of the tiny effects that small masses create, deserves serious celebration before we contemplate the frustration that this method brings.[2]

Astronomers detected their first exoplanet microlensing event in 1995, using a network of small telescopes that were judged obso-lete for most astronomical research. This success naturally spurred increased efforts, resulting in ongoing surveys that monitor mil-lions of stars and search for the increases in starlight that signal the occurrence of microlensing. Because a typical rise in stellar brightness occurs over a few weeks, the astronomers involved in these searches reduce their data on a daily basis, and they alert one another to concentrate their observations toward a particular star, looking for the brief additional increase in brightness that an exoplanet will generate. With telescopes located around the world, these searches for microlensing have lived up to the proud boast of Scott Gaudi, one of its chief practitioners, that "the sun never rose on our collaboration" (though of course clouds can always create a problem).[3] The duration of the rise and fall in brightness during a microlensing event varies in proportion to the square root of the mass of the object producing the effect. Thus, for ex-ample, if the star-induced event lasts for a month, the subsidiary event from a planet with one one-thousandth of the star's mass will last for a day.

Microlensing searches for exoplanets typically direct their atten-tion to stars that lie along the line of sight toward the center of our Milky Way galaxy, where the density of stars rises highest and offers the greatest prospects for success. These surveys, which have examined hundreds of millions of stars that typically lie tens of thousands of light years from the solar system, have now revealed about 70 exoplanets, a number sufficient to allow astronomers to draw some overall conclusions. Although the planets' masses and distances from their stars affect the probability of detecting them, the sweet spot for the most easily detectable microlensing events belongs to planets that orbit at about three times the Earth–sun distance from their stars.

Throughout the past three decades, astronomers have steadily refined their microlensing search techniques, using Earth-based observatories in combination with the Spitzer Space Telescope described in Chapter 9. The ground-based observatories sample comparatively wide regions of the sky, a couple of degrees on each side, on timescales that range from 15 minutes to a couple of hours, seeking to catch the start of a typical star-induced gravitational-lensing event, which lasts for several days. If the star has a modest-sized planet, the much smaller microlensing from the planet takes only a few hours. The astronomers have developed a system that avoids the usual multiweek planning for observations with the Spitzer telescope. This system allows them to direct the spacecraft toward a gravitational-lensing event within a couple of days, giving them a good chance to observe from space, as well as from the ground, the blip that a planet's gravity creates.

In at least one case, dating from 2012, the technique discerned two planets orbiting a single star. In this event, labeled as OGLE-2012-BLG-0026, astronomers detected microlensing around a star similar to the sun, 13,000 light years away, that showed planets reminiscent of Saturn and Jupiter: One of them, slightly closer to its star than Jupiter is to the sun, has a mass about half of Saturn's, while the other, orbiting almost as far from its star as Jupiter does from the sun, has a mass seven-eighths of Jupiter's. These planet–star distances represent lower limits, because the microlensing event may not have occurred at a time when the separations between the star and its planets reached their maximum values.

The down side of searching for planets with microlensing resides in the fact that a planet will appear in the data once and only once, providing no chance for any follow-up investigation, because the chance line-up that signals the existence of a particular planet will never recur on any human timescale. Some disappointment inevitably arises from the realization that no matter how intriguing a planet's mass and distance from its star may prove to be, we have no chance to study the system any further. Astronomers have nobly surmounted this frustration, as they are aware that the microlensing technique continues to offer an excellent means of sampling the stars in the Milky Way for planetary companions.

Microlensing has already shown that planets exist in abundance not only in our own neighborhood, where other techniques work best, but throughout our galaxy as well. The 70 planets detected so far by microlensing help to fill in an observational gap, because this technique can find planets considerably farther from their stars than most of the stars found by the radial-velocity and transit observational techniques that have yielded exoplanets in far greater numbers. Future observations during Kepler's K2 mission should reveal many more exoplanets through microlensing. In particular, the astronomer Calen Henderson notes that "K2's orbit is in the Goldilocks zone" to find exoplanets in situations when a star with roughly half the sun's mass and a distance of about 12,000 light years creates gravitational lensing of one of the stars in the central bulge of the Milky Way, about twice as far away.[4]

The Earthlike Planet
OGLE-2016-BLG-1195Lb

Microlensing offers the possibility of finding planets with masses comparable to Earth's. The current record holder for a low-mass exoplanet found through microlensing appeared in June 2016, and carries the designation OGLE 2016-BLG-1195Lb. (OGLE stands for the Optical Gravitational Lensing Experiment, a worldwide observing project based at the Warsaw Observatory; 2016 shows the year of observation; BLG denotes the bulge of the Milky Way; the number specifies the event; the uppercase letter indicates the star; and the "b," as usual, refers to the planet.) Astronomers followed this microlensing event with three different systems: the OGLE network, which made the initial detection, the KMTNet (Korea Microlensing Telescope Network), and the Spitzer Space Telescope. Like the sun, OGLE-2016-BLG-1195Lb and its star (which carries the same designation without the "b") lie within the disk of the Milky Way, in this case 13,000 light years from Earth. The microlensing observations showed that the star has only about 8 percent of the sun's mass, and the planet, whose

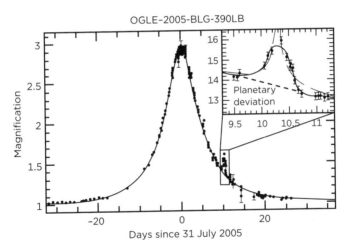

OGLE-2005-BLG-390LB

Figure 10 This graph records the observations of the gravitational lensing event, designated as OGLE-2005-BLG-390LB, that were made over eight weeks in 2005, initially by only one telescope, then with several others added after July 23. Superimposed on the impressive magnification of the light from a distant star caused by a closer star's gravitational lensing is a much smaller "bump," which appeared around August 10 and is shown in magnified form at the upper right. The bump arose from a planet with a mass calculated at 5.5 Earth masses, and an orbital radius of 2 to 4 AU. (Courtesy of Jean-Philippe Beaulieu)

microlensing event lasted for about 2.5 hours, has approximately 42 one-millionths of the star's mass, or about 1.43 times the mass of our own planet.[5] This super-Earth orbits its star at about twice the Earth–sun distance. Because its star generates less than one percent of the energy per second that our sun does, the exoplanet's temperature falls far below the coldest temperature on Earth.

For those who seek Earth's twin, OGLE-2016-BLG-1195Lb offers good news and bad news. On the one hand, the planet has a mass close to the Earth's, and its orbit resembles our own in size. On the other hand, its star has only about 8 percent of the sun's mass, making it an ultracool red dwarf similar to the fascinating red dwarf called TRAPPIST-1 (see Chapter 9). But on the bright side once again, the fact that both TRAPPIST-1 and OGLE-2016-BLG-1195L, each with 8 percent of the sun's mass, have planets—discovered by entirely different techniques—implies that planets may well be common around ultracool red dwarf stars. In fact, the low masses of these two stars place them almost precisely on the borderline between a star and a brown dwarf, which cannot liberate energy through nuclear fusion and radiates only the heat stored from its formation era. In either case, however, the star's impressively low luminosity renders its newfound planet an "iceworld," reminiscent of the future state of Earth after our sun becomes a white dwarf, 7 or 8 billion years from now.

The current record for the most massive planet found by microlensing currently belongs to an object with the designation OGLE-2011-BLG-0402Lb, detected in 2013. This planet, with roughly 9 times Jupiter's mass, orbits at a distance of 0.2 AU from a star with $\frac{1}{40}$ of the sun's mass, 6,500 light years from the solar system. In other words, here we have a super-duper Jupiter in a half-Mercury-sized orbit around a very low-luminosity star, detected once and only once with Einstein's lens.

Measuring the Mass of a Dwarf with Einstein's Relativity

In 2017, astronomers announced that they had used the effect that Einstein described in his general relativity theory to measure the mass of a white dwarf, an object that packs the mass of a star into a volume comparable to Earth's. A white dwarf represents the shrunken core of a once-active star that has lost its outer layers—a fate that lies in store for our sun once the sun can no longer use nuclear fusion to create kinetic energy. At intervals over a six-year period, a team of astronomers led by Kailash Sahu of the Space Telescope Science Institute used the Hubble Space Telescope to study the sixth-closest white dwarf to the sun, Stein 2051 B.[6] As the white dwarf moved across the line of sight toward a more distant star, located almost directly behind the white dwarf, the astronomers could measure the changing amounts of light that the white dwarf deflected from the star. The application of Einstein's theory of how gravity bends light then showed that Stein 2051 B has a mass about 68 percent of the sun's.

Buoyed by their recent success in using microlensing to detect exoplanets, the astronomers engaged in this effort have turned their attention (using the time that they are granted to use the Hubble Space Telescope) toward Proxima Centauri, the red dwarf star closest to the sun and possessor of its recently found exoplanet, Proxima b. Determining the mass of this star will allow a more accurate measurement of the mass of its planet (see Chapter 4). Because Proxima Centauri has a much larger size and a much smaller mass than Stein 2051 B does, detecting and measuring the deflection created by the star's gravitational force poses a much greater challenge than the already impressive detection of the same effect from a white dwarf. Einstein showed that the amount of deflection produced by an object depends on the object's mass divided by its radius. This mass-over-radius ratio favors Stein 2015 B, which has 5.5 times Proxima Centauri's mass but only $\frac{1}{10}$ of its size, by a factor of 55. The team using the Hubble Space Telescope must therefore seek to measure impressively small angles of the deflection of light rays. If they succeed, they will have achieved the first use of the gravitational deflection of light to measure the mass of an exoplanet detected by other means.

8

.

TWO MINOR METHODS FOR FINDING EXOPLANETS

As we described in previous chapters, astronomers have found exoplanets with four major techniques—transit and radial-velocity observations, gravitational lensing, and direct imaging. They have also used a fifth method, the exact timing of pulsar radio emission, to discover a few unusual and anomalous pulsar planets. In the interest of completeness, we should now add four more search methods, two of which have so far proven unsuccessful, while the other two—orbital brightness modulation and starlight polarization—have actually provided us with a few more exoplanets. This will bring the number of successful techniques to seven, with at least two more awaiting the crown of consummation.

Radio Observations: A Fine Idea Fails

Radio astronomy, which sprang quickly from technological advances introduced during the Second World War, has added immensely to our store of astronomical knowledge. A variety of objects and events, ranging from the earliest moments of the universe to the most extreme individual explosions and weirdest astronomical situations, have created the long-wavelength radiation that has allowed radio astronomers to peer into the heart of the cosmos. On a more mundane level, many of the sun's planets, most notably Jupiter, as well as many active stars, radiate strongly in the radio domain, typically with transitory intense bursts of radio waves. If Jupiter-like planets around stars close to the sun emit as much of this radiation as our Jupiter does, radio observations offer the chance to find them directly, without recourse to subtle approaches such as the radial-velocity and transit techniques. However, because specialized large arrays of radio dishes would be required to achieve these detections, a lack of funding has so far prevented this approach from bearing fruit.

Another approach in applying radio-astronomy techniques to the search for exoplanets has been under way for years, but likewise without success. If the Milky Way contains civilizations whose stage of development roughly mimics our own, they, too, might employ radio waves for their localized radio and television communications, for radar systems, and even, as we have occasionally done, to signal their existence to the cosmos. Although funding for SETI, the generic term for the search for extraterrestrial intelligence, has often been scarce (or, in the case of efforts financed by the United States government, eliminated by statute in 1994 and never again proposed)[1], SETI-oriented astronomers have completed a variety of investigations that either surveyed the sky or directed their attention toward particular stars. Over the course of more than four years, one of the largest of these enterprises used the several dozen radio dishes in the Allen Telescope Array, located at the Hat Creek Observatory in northern California, to hunt for narrow-band radio signals arriving from the directions of more than 9,000 stars, including more than 2,000 stars in the Kepler catalog.[2] As one would surmise from this project's lack of publicity, this search found nothing suggestive of a civilization employing radio waves as we do. Let us, therefore, turn our attention to two methods that have provided low-key, modestly successful approaches to the detection of exoplanets.

Orbital Brightness Modulation

As astronomers steadily improved their ability to find exoplanets through their transits, they recognized a useful sidelight to their most successful detection method. Transit observations can reveal only the small minority of planetary systems whose orientation carries one or more planets directly across our line of sight to a star. But what happens if a planet's orbit takes it nearly, but not exactly, along this path?

In most cases, nothing detectable will occur. However, if a large planet orbits close to its star, the amount of starlight reflected in our direction by the planet will change in synchrony with the planet's orbital motion. Even though the planet's reflected light adds only a tiny amount to the stellar output, its contribution may prove detectable because it varies periodically as the planet moves in orbit. When the planet passes almost directly between our line of sight and its star, that contribution falls to a minimum; half an orbit later, it reaches its maximum value. Searches for planets that monitor the brightnesses of stars to detect planetary transits can also disclose nearly transiting planets—if they have close-in orbits, are sufficiently large, and reflect most of the starlight that reaches them.

Planetary detection by this "orbital brightness modulation" approach has furnished only half a dozen entries in the thousands-long list of exoplanets. The prime examples of these, Kepler-70 b and Kepler-70 c, orbit the star Kepler-70, about 4,000 light years away. As could be expected, with the exception of the planets detected around pulsars, these two planets rank among the closest to their parent stars, orbiting at 0.0060 and 0.0076 AU. Kepler-70 b completes each orbit in 346 minutes (thus more than four times each day), while Kepler-70 c takes 494 minutes for each orbit. The planets have minimum masses equal to 44 percent and 66 percent of the Earth's, and their diameters are estimated at 76 percent and 87 percent of our planet's. Their reflected light has thus shown them to be rather incredible objects: Earthlike planets orbiting their star at a distance less than $\frac{1}{100}$ AU.

Kepler-70, far from being a sunlike star, is a "subdwarf," a star that has already passed through its red-giant phase (see Chapter 9), has shed its rarefied outer atmosphere, and is now slowly contracting and cooling toward its eventual white-dwarf phase. With a surface temperature almost five times the sun's, Kepler-70 radiates about 23 times more energy per second than the sun does. As we might expect from the laws of physics, the star's two known planets have surface temperatures at the top end of the exoplanet scale. Indeed, Kepler-70 b, the inner planet, which receives almost a million times more energy per second from its star than we do from the sun, has a surface temperature estimated at 7,150 K— hotter than the surface of the sun itself![3]

To endure such temperatures, and to have survived the heightened output from the star's previous red-giant phase, these planets must be tough. Kepler-70 b and Kepler-70 c each consist of an Earth-mass amount of material capable of surviving at temperatures of many thousands of degrees. We may reasonably speculate that these planets consist largely of iron and other metals, such as nickel and zinc. Astronomers must attempt to explain how these planets, like many of those found via the radial-velocity method, have come to orbit so close to their stars: Did they form at these comparatively tiny distances, or migrate inward after they had formed?

Starlight Polarization

For completeness, we should cite a final successful method (final for now, at any rate) of planet detection, one that employs the polarization of light. We may think of light waves as made of massless particles, called photons, that can travel through empty space, vibrating as they go in directions perpendicular to their direction of travel. (Sound waves, by contrast, consist of alternate zones of compression and rarefaction along their direction of travel through a medium such as air or water.) The perpendicular vibrations of light can be described in terms of two mutually exclusive components, such as "up and down" versus "side to side," or "vibrating clockwise" versus "vibrating counterclockwise" (future generations will surely wonder what "clockwise" and its opposite may have meant, but for now the terms usually provoke recognition). If we generate light with only one polarization component—up and down, for instance, but not side to side, we can also create polarization filters that will either allow or block that component, depending on our goal. Edwin Land, the founder of the Polaroid Corporation, manufactured the first commercially successful polarization filters; he applied his innovation to create better sunglasses and (sadly, without success within the automobile industry) headlights that would emit polarized light and complementary filters that would reduce the glare from the headlights of oncoming cars. Today astronomers routinely employ polarization as one of their tools for analyzing the radiation from distant objects: Ordinary starlight shows little or none of it, whereas the radiation from complex objects, such as stellar masers and radio galaxies, often arrives with significant polarization. Even the universal cosmic background radiation, left over from the first few minutes following the big bang that began the universe, shows some polarization.

When a planetary atmosphere reflects the light from its star, it often increases the degree of the starlight's polarization, in an amount that depends on the angle at which the starlight strikes the planet, as seen by a particular observer. In theory, this effect could be used to detect planets by analyzing the changing amount of polarization in the combined light of the star and of the much smaller amount of light reflected from a planet. In practice, this has not yet occurred for actual discoveries, although in the case of the star HD 189733 and its exoplanet (HD 189733 b), astronomers have observed this polarization effect, the result of a planet that had already been detected by the transit method. HD 189733 b, only about 63 light years away, draws attention for other reasons: It ranks as the closest hot Jupiter that undergoes a transit, the first planet to have its transit detected in x-rays, and the first whose atmosphere has been found to contain a thin haze-layer rich in microscopic dust particles.[4]

Although orbital brightness modulation and planet-induced polarization offer specialized avenues toward a better understanding of some exoplanets, most of us will remain content with this brief summary of their role in planet detection. Before we examine a far more fascinating aspect of exoplanets—the possibility that some of them teem with life—we ought to take a tour through a "rogues gallery" of the most intriguing planets detected so far.

9

·

A GALLERY OF STRANGE
NEW PLANETS

The broad conclusions that we may draw from nearly 4,000 exoplanets, which we will summarize in Chapter 12, primarily depend on the statistical characterization of these new worlds. This makes good sense: Science proceeds by using statistical analysis to separate reliable results from "anecdata." We would do well, however, to delineate the outer limits of the characteristics of currently known exoplanets by examining some of the most bizarre and most interesting planets and planetary systems that different observational techniques have revealed. The criteria for inclusion on the list of these outstanding planets must, of course, remain subjective, but they rest on the notion that "most bizarre and most interesting" corresponds to "least expected" and "most unlike anything found in the solar system."

Some Outstanding Kepler Planets

Many of the weirdest members of the exoplanetary zoo proudly bear their informal designation as "Kepler planets," which is hardly a surprise since the Kepler spacecraft's observations of their transits have provided more than half of the current entries in exoplanet catalogs. Among the plethora of Kepler planets, half a dozen stand out as particularly remarkable. A tour through these exoplanetary standouts will underscore the wide variety of planets that speckle our galaxy, and, in some cases, will raise expectations that astronomers may, before long, verify that at least some exoplanets provide Earthlike conditions, favorable to the origin and evolution of life.

Tabby's Star: Probably Not an Alien Civilization

One of the strangest on this list of strange exoplanets would in fact belong to a different list, should it prove to be no planet at all, as some have suggested. The weirdness in this case surrounds KIC 8462852, a Kepler star that carries the popular names of "Tabby's Star," "Boyajian's Star," and the "WTF Star." The first two designations reference Tabetha Boyajian, the astronomer at Louisiana State University who led studies of this star's variable brightness. The abbreviation "WTF" is said to stand for the phrase "Where's the Flux?," the subtitle of the report on those studies that appeared in 2016. For those who care, the KIC prefix stands for "Kepler Input Catalog."[1]

Boyajian and her coinvestigators found that KIC 8462852 exhibits large and sudden decreases in its brightness at apparently random times. Many young stars undergo such changes, which are associated with their "struggle" to achieve a near-perfect balance between the outward flow of energy released in their nuclear-fusing cores and their tendency to collapse under their self-gravitational forces. But Tabby's Star, as we may call it in honor of Professor Boyajian, does not belong to the young-star category. The star is located about 1,275 light years away from us in the direction of the constellation Cygnus; its spectrum places it firmly among the majority of mature, "main-sequence" stars similar to our sun, though Tabby's Star has about 40 percent more mass than the sun and 4.7 times its luminosity.

The decreases in this star's brightness, observed over a four-year period, fall roughly into two categories—small and large—with none of them showing the periodic behavior that arises when one or more orbiting planets transit the star, or when older stars pulsate with metronomic regularity. The numerous small dips in brightness might well arise from a swarm of small objects—something like the comets that surround the sun but orbiting this star at much lesser distances—that would decrease the star's observed brightness at random intervals. But the two observed large decreases, by about 15 percent in March 2011 and 22 percent in February 2013, could hardly result from either a large planet or a host of comets. These events lasted for about a day, significantly longer than the brightness decreases that a planet would produce.

Another anomalous aspect of the star's brightness variations emerged from the wealth of data accumulated by the Kepler spacecraft. Through most of the four-year run of observations, the star grew dimmer at a rate of about 0.34 percent per year, but in the midst of this general behavior, it became about 2.5 percent dimmer within 200 days, and then it resumed its original, much slower, rate of dimming. Since the Kepler data offered hundreds of comparison stars in the same general direction and at similar distances, astronomers could determine that Tabby's Star's behaved uniquely: No other star showed a similar record in the "light curve" that shows how a star's brightness varies over time.

What might explain this impressively anomalous behavior? Because astronomers understand the behavior of mature stars so well and have thousands upon thousands of examples that support their understanding, they nearly completely reject the hypothesis that Tabby's Star itself can explain the observed variations in the light received from its vicinity. Instead, something in the star's surroundings must be the culprit—but what? Comets might be falling inward toward the star in great numbers, most especially during the periods of the two deepest dips. Dustlike material orbiting the star—typical of stars as they form, but absent from the vicinity of mature stars—could swirl in complex ways, blocking different amounts of light at different times.[2] Rings of material in the far outer solar system might provide the answer, if those rings happened to lie between Kepler's line of sight and the star.[3] Perhaps the data have errors, either in the measured light variations or in the spectral observations that establish Tabby's Star as no youngster.

Or perhaps—and here the public mind grows most heavily engaged—a megastructure surrounds the star, a variation on a "Dyson sphere," the enormous light- and heat-trapping shield around a star that the physicist Freeman Dyson suggested in 1960 could result from an advanced civilization's attempt to capture all of its star's energy. If the shield were only partial, perhaps composed of many parts in orbit around the star, its motions could at times block a large portion of the stellar output from our view.[4]

In scientific circles, all hypotheses remain subject to a basic approach: How can we attempt to check on their validity? If no reasonable answer exists, the hypothesis will likely be set aside as intriguing but not useful. If methods do exist to validate or reject a hypothesis, it has already passed one part of the crucial test, and faces only the remaining part: an experiment that will allow scientists to check its results against the original suggestion.

The hypothesis that great clouds of dust surround a star faces a useful and determinative test. Since dust particles absorb starlight, the star's energy will heat them so that they emit infrared radiation. Studies with NASA's Infrared Telescope Facility on the summit of Mauna Kea have failed to find any "infrared excess" from large numbers of dust particles within a few astronomical units of the star. A search for radio signals from the star's vicinity that could imply the existence of a civilization roughly similar to ours also failed to yield positive results.[5] For now, we may fairly say that we lack a good explanation for the remarkable behavior of Tabby's Star.[6] Further monitoring of the star, partially funded by a Kickstarter campaign that Boyajian led, may provide additional data that will lead to a better understanding of this cosmic anomaly.[7]

Kepler 10 b: An Earthlike Planet Close to Its Star

Despite its number, Kepler-10 b was the first Kepler planet to be found, verified with observations made at the Keck Observatory in Hawaii. This exoplanet orbits a star that resembles the sun, but the star has an estimated age of 10 billion years rather than the sun's 5 billion. Kepler-10 b's transits reduce its star's light by 1 part in 10,000 every 20 hours, indicating that the planet has a diameter about 50 percent greater than Earth's and an orbital distance of only 0.017 AU. Radial-velocity observations of Kepler-10, about 560 light years away, confirmed this orbit and showed that the planet's mass equals 3.33 times the Earth's, giving Kepler-10 b an average density of 5.8 grams per cubic centimeter, slightly larger than the Earth's 5.51. (Recall that for all transiting planets, no uncertainty exists in the mass derived from radial-velocity measurements, because we know that the planet's orbit aligns precisely with our line of sight.) Further observations found the more modest diminution in the light from the star–planet system that occurs when Kepler-10 b passes directly behind its star.

Using a host of techniques too subtle to be described briefly, astronomers deduced the existence of a second planet, Kepler-10 c, in orbit around this star, at a distance 14 times greater than Kepler-10 b's, or 0.24 AU. Kepler-10 c, a comparatively giant planet, has 17 times the Earth's mass and 2.35 times its diameter, but takes just 45 days for each orbit. This example among the first

Kepler planets emphasizes how widely exoplanet systems may differ from the model suggested by the solar system, with a rocky, Earth-sized planet extremely close to its star, along with a much larger, more massive but also rocky planet at a distance comparable to Mercury's distance from the sun.[8]

Kepler-16 b, Familiarly Known as Tatooine

In 2011, the Kepler astronomers reported the first exoplanet found to orbit not one but two stars. Kepler-16, about 195 light years away, turned out to be a binary-star system whose two stars, each smaller and less luminous than the sun, maintain a distance of 0.22 AU from each other while they perform 41-day orbits around their common center of mass. Well outside the stars' mutual orbits, a giant planet with 75 percent of Jupiter's diameter and a mass comparable to Saturn's orbits the center of mass, following an almost circular trajectory with 70 percent of the Earth's orbital diameter as it takes 229 days for each orbit. Kepler-16 b's orbital size and period closely match Venus's orbital parameters, reflecting the fact that the sum of the masses of its two stars, 69 and 20 percent of the sun's, come close to the sun's mass, so a planet orbiting their center of mass experiences a gravitational force similar to the force that the sun exerts on its planets at the same distance.[9]

In the Kepler-16 system, the planet's orbit and the two stars' orbits around their common center of mass align with our line of sight, so the Kepler astronomers could detect the stars eclipsing one another in each orbit. In one eclipse, the smaller and fainter star covers only a portion of the larger one, decreasing the system's total brightness by 13 percent. During the other eclipse, when the larger star covers the smaller one completely, the observed brightness declines by only 1.6 percent, because the larger star's luminosity far exceeds the smaller one's. Of course, these declines rank as enormous in comparison with the brightness decreases that occur in a planetary transit, often measured in parts per hundred thousand.

The alignment of the binary stars' orbital plane with the plane of the planet's orbit corresponds to astronomers' theories of how stars and planets form within a rotating disk of gas and dust. Kepler-16 b orbits at 0.7 AU from the system's center of mass, while the two stars have distances of 0.05 and 0.17 AU from this center. Previous studies had implied that a planet as close to a double-star system as Kepler-16 b could not remain in the same orbit for millions of years. A closer analysis, however, concluded that the planet can indeed maintain its orbit for longer periods of time, because it finds itself in a "resonance cell" that keeps it just outside the orbital "chaos zone" around the two stars. The term "resonance" refers to an integer or half-integer ratio between two orbital periods, in this case the times that the planet and the binary-star system take to orbit the center of mass of the system. In a dazzling display of the celestial-dynamics calculations that distinguish professionals from amateurs, the Russian astrophysicists Elena Popova and Ivan Shevchenko showed that if this ratio were to equal 5:1 or 6:1, the planet's orbit would be unstable, so the planet would soon follow quite a different path, but that a ratio of 11:2, halfway between the integer values, leaves Kepler-16 b just within the zone of orbital stability.[10]

Kepler-22 b: A Planet with Earthlike Temperatures

The star Kepler-22, 600 light years from us in the direction of the constellation Cygnus, qualifies as the sun's near sister, with 97 percent of the sun's mass, 98 percent of its diameter, and a surface temperature a few hundred degrees less than the sun's 5,778 K. (For those wedded to the Fahrenheit scale, the sun's surface temperature corresponds to just under 10,000 F, which provides one of the easiest-to-memorize solar-system descriptors.)

At the end of 2011, the Kepler spacecraft found a planet around this star that turned out to be the first transiting planet known to orbit a sunlike star at a distance similar to the Earth–sun distance. Kepler-22 b follows a nearly circular path at a distance of 0.85 AU from its star, taking 289 days for each orbit. Because Kepler-22 generates somewhat less energy per second than the sun does, even the planet's comparatively smaller distance from its star implies that Kepler-22's average surface temperature, if the planet has no atmosphere to trap heat, should be about 262 K, equal to 11 degrees below zero on the Celsius scale. The planet's equatorial regions, however, should have temperatures above the freezing point of water, so that Kepler-22 b offers slightly more favorable conditions than Mars does for the possible existence of life on its surface (see Chapter 12).[11]

Kepler-22 b's transit revealed the planet's diameter, about 2.04 times the Earth's. This fact makes Kepler-22 b what astronomers now call a "sub-Neptune," a type of planet different from anything in the solar system: too large to be—most astronomers believe—a rocky world like the sun's four inner planets, yet too small to resemble the four gas giants of the outer solar system, each of which has at least 3.9 times the Earth's diameter.

Sadly, we lack any good estimate of the planet's mass. If Kepler-22 b orbited much closer to its star, the radial-velocity technique would stand an excellent chance of providing this missing data point. But a planet much less massive than Saturn, orbiting at an Earthlike distance from its star, lies outside astronomers' capabilities, at least for now. At best, astronomers can tell us only that measurements of Kepler-22 show that its planet has a mass less than about 35 Earth masses, hardly a surprise for a planet with about eight times the Earth's volume.

If we assume, as seems reasonable, that the orbits of any other planets around Kepler-22 would lie in the same plane as Kepler-22 b's, the absence of transits other than Kepler-22 b's would imply that the star has no planet at least as large as Venus or Earth out to distances of about 2 AU. At larger distances, a planet's transits would occur so rarely that the Kepler spacecraft might have had no opportunity to observe even a single one. Kepler-22 b must therefore be king of the star's inner planetary region. The planet is probably too large to be made of rock and metal, but too small to consist mainly of gas, as the sun's giant planets do. Could Kepler-22 b turn out to be a "waterworld," with liquid water providing a significant portion of the planet's mass? (We should always recognize that although the oceans cover 70 percent of the Earth's surface, their depth of only a few kilometers makes them a tiny bubble on the overwhelmingly solid planet on which we live.) For now, this remains intriguing speculation, and Kepler-22 b's composition and true nature remain unknown.[12]

Kepler 36: Two Planets in Close Proximity

Among the many planets found by the Kepler spacecraft, the two-planet system of Kepler-36 has a claim to extreme weirdness, not for the characteristics of its individual planets, but for the planets' mutual relationship. The star itself deviates from the normal run of sunlike stars. Fifteen hundred light years away in the direction of the constellation Cygnus, Kepler-36 has a mass 7 percent larger than the sun's and a surface temperature several hundred degrees hotter than the sun's. However, the star's radius equals 1.63 times the sun's, which places Kepler-36 in the category of "subgiant stars." Subgiants have reached the end of their nuclear-fusing lifetimes, the eras when they steadily fuse hydrogen nuclei into helium within their central cores. Eventually, as a star exhausts its supply of hydrogen nuclei in its center, it will contract and heat its interior to the point that hydrogen fusion can continue in a shell around the core. Before long (that is, within millions rather than billions of years), the star's rate of nuclear fusion will increase significantly. Part of the additional energy released by fusion will escape, while the remainder will make the overlying layers of the star expand. Eventually, the star will become a "red giant," with a highly rarefied outer layer surrounding its nuclear-fusing region.

Kepler-36 now occupies the initial stages of this evolutionary path. Its two known planets, Kepler-36 b and Kepler-36 c, were presumably created at the same time as their star, though whether they have followed their present orbits for most or all of their lifetimes remains unknown. The two planets draw attention for their proximity: They orbit their star at distances of 0.115 and 0.128 AU, leaving a gap of 0.013 AU—a bit less than 2 million kilometers—between them. Since both orbits are nearly circular, the planets never come closer to one another than this distance. Their diameters, measured by the planetary transits' effects on starlight, equal 1.5 and 4.0 times the Earth's, so this star has two planets, each significantly larger than our own, that orbit their star at one-ninth and one-eighth of an AU, approaching one another to within one-tenth of that distance.[13]

The two planets' close approaches cause them to interact with one another gravitationally to a significant degree. These interactions induce subtle changes in the planets' motions, which in turn cause variations in the precise times of their transits across the star's disk. Measurement of these changes allows the determination of the planets' masses, which turn out to equal 4.5 and 8.1 times the Earth's. Combined with the planets' diameters, these masses imply a notable divergence in the planets' average densities: 7.4 times the density of water for Kepler-36 b, but only 70 percent of water's density for Kepler-36 c, whose volume exceeds its close neighbor's by a factor of 19. We may reasonably conclude that although these planets' orbits may nearly match, one of them must be a rocky planet, with an average density one-third larger than the Earth's (5.5 grams per cubic centimeter), while the other, with an average density less than water's, seems to be a fluffed-up mini-Saturn. The next chapter places these planets in the categories of "super-Earths" and "sub-Neptunes."

If other transiting planets exist around Kepler-36, they lie so far from their star that each of their orbits would take two years or more—that is, they move at distances more than 15 times those at which the two known planets orbit. We may someday obtain an explanation of how these two exoplanets, so similar in their trajectories but so noticeably different in their overall characteristics, came into existence (wherever that may have been in their circumstellar surroundings) and acquired orbits so close to each other. For now, we can admire one more oddity in our list of intriguing planetary systems.

Kepler-452 b: An Earthlike Planet?

The quest for Earth's twin, or at least its close cousin, has driven much of the enthusiasm in exoplanet searches.[14] The list of primary qualifications for the long-sought title of Earth's twin begins with a planet's Earthlike mass and diameter, includes its motion in an orbit similar to Earth's around a sunlike star, and culminates with an Earthlike atmosphere. Kepler-452 b fulfills the orbital and stellar requirements, but it falls short in the planet's size and mass.

Kepler-452, 1,400 light years away, resembles the sun in its mass, size, surface temperature, and luminosity. A bit richer than the sun in elements heavier than helium, Kepler-452 also surpasses the sun in its age, estimated at 6 billion years. In 2015, the Kepler scientists announced that they had found a single planet around the star, orbiting at a distance 5 percent greater than the Earth–sun distance and taking 385 days for each orbit. If Kepler-452 b had the same size and mass as Earth, the publicity machine that promotes astronomical news (not quite as great as it should be) would have placed this planet near the top of the list of objects on which we should search for life on other worlds.

In fact, Kepler-452 b has a diameter close to 1.55 times the Earth's, with an estimated error of +30 or −20 percent. This size implies that the planet has a volume about 3.75 times the Earth's, so if it has a rocky composition similar to our own planet's, then its mass would likewise be about 3.75 times larger.[15] Despite its "failure" to be a duplicate Earth, Kepler-452 b is indeed an exoplanet whose orbit around a sunlike star gives it an Earthlike temperature, with a likely rocky surface that provides more than double the Earth's surface area.[16] The researchers at the SETI Institute promptly directed the Allen Telescope Array in northern California to start searching in the direction of this star and its planet for any radio signals possibly generated by nonnatural processes. The fact that this effort remains largely unknown to the public signals that this search has yet to achieve success.

Exoplanet Transits Observed from Ground-Based Observatories

Although the Kepler spacecraft's observations have provided the vast bulk of exoplanets found by the transit method, this technique began here on Earth. As we described in Chapter 5, the atmosphere's ever-changing refraction limits the accuracy of astronomers' measurements, so that they can determine stars' brightnesses only to within a few tenths of 1 percent. This restriction nevertheless allows the detection of planetary transits that change their stars' apparent brightnesses by amounts greater than this limit. When applied to the wide range of planetary diameters, this restriction implies that ground-based transit searches can find planets the size of Jupiter or Saturn, or perhaps somewhat smaller, around sunlike stars, and—in a tribute to the astronomical usefulness of a star's small size—these searches can find Earth-sized planets, or even smaller ones, that orbit M stars. Several searches have brought the former possibility to success, and another one temporarily carried exoplanets around M dwarfs into the center of public attention.

Osiris, a Mighty Weird Planet

We have previously met the happy result that occurs when astronomers can detect a transiting exoplanet through the radial-velocity changes that the planet induces in its star, which allows them to measure the planet's mass directly. How many exoplanets belong to this category of high-caliber observations? As the number undergoes rapid expansion, let us take a look at the first planet observed during its transit across a star. This object carries the astronomical designation HD 209458 b, but its popular name rings a far sweeter bell: Osiris! To ancient Egyptians, the god Osiris governed transition and regeneration, judging the dead and overseeing the Nile's annual flood.

Osiris, as we will also call the planet to signal an intimate acquaintance, has a claim to the twin titles of "best studied" and "weirdest" among all the exoplanets. The star lies about 150 light years from the solar system in the direction of the constellation Pegasus, and thus has about three times the distance to 51 Pegasi, which hosts the first nonpulsar exoplanet to be discovered. Selected for its resemblance to the sun in size, mass, and luminosity, HD 209458 drew close scrutiny from the two groups that had found the first radial-velocity planets in 1995 and 1996 and had continued their (usually) friendly rivalry during the following years. In 1999, both groups announced that they had found a close-in planet around HD 209458, which completes a nearly circular orbit every 3.5 days at only 0.047 AU from its star.

Aware of the fact that planets orbiting closer to their stars are more likely to undergo a transit than planets orbiting at greater distances, two other teams of astronomers then sought evidence of a transit and found it: the first observed transit of an exoplanet, achieved almost a decade before the launch of the Kepler spacecraft. Because Osiris's transits cause a comparatively hefty reduction in the star's brightness, they could be observed from ground-based observatories without requiring the much better clarity and stability provided by space-based observatories. Each transit of Osiris takes approximately three hours and reduces the star's brightness by about 1.5 percent. Because HD 209458 has approximately the same size as the sun, but a radius about 14 percent larger, and because a transit of Jupiter, if seen by a distant observer, would reduce the sun's brightness by 1 percent, the conclusion follows that Osiris must be notably larger than Jupiter. Osiris's diameter in fact equals 1.4 times Jupiter's and 16 times the Earth's, making the planet a "hot Jupiter" whose mass, estimated from the exoplanet's infrared spectrum as described in the next section, equals about 70 percent of Jupiter's and 220 times the Earth's.[17]

Radial-velocity measurements later established that Osiris has a mass equal to 0.63 Jupiter masses multiplied by the sine of the exoplanet's orbital inclination (see Chapter 4). However, because Osiris transits its star, we know that the orbital inclination equals 90 degrees, so its actual mass equals the full 0.63 times Jupiter's mass.

The Spitzer Space Telescope Makes Infrared Observations

Osiris's greater astronomical fame sprang from infrared observations that covered a wide spectral range, including infrared studies of the planet and its star. Rendered impossible from the Earth's surface because of the absorption of infrared by our atmosphere, such observations form the heart and soul of NASA's great infrared space observatory, formerly known as the Space Infrared Telescope Facility. Now renamed after the distinguished astrophysicist Lyman Spitzer, the Spitzer Space Telescope has studied infrared radiation from cosmic sources since 2003, operating almost a decade longer than its planned lifetime. To escape the infrared glow from our planet, the telescope follows an Earth-trailing orbit around the sun (similar to that later adopted by the Kepler spacecraft) along a path much like the Earth's, but one that carries it farther from our planet by about 15 million kilometers each year.

The telescope observes the cosmos without terrestrial interference, but its tendency to create its own infrared radiation would nevertheless frustrate the achievement of its astronomical goals. To counter this problem, the telescope's designers used liquid helium to cool its 85-centimeter mirror to a temperature just 5.5 K above absolute zero. This allowed the telescope's three observational instruments to function as planned for almost six years, while the helium coolant slowly evaporated. Once its cooling function vanished in 2009, the telescope warmed to the ambient space temperature, still maintaining some of its observational capability: Two of its three instruments, which remain comparatively unaffected by the telescope's heat, operate at shorter infrared wavelengths. Astronomers have recently used these instruments to obtain direct infrared images of planets in orbit around nearby stars (see Chapter 6).

For observations made within the modest range of infrared wavelengths accessible from Earthborne observatories, the Spitzer telescope operates at a disadvantage, because its mirror spans only about one-tenth of the diameter of either of the Gemini telescopes, and therefore it has only about $1/100$ of the telescopes' ability to gather infrared radiation. This limits the Spitzer telescope's ability to observe faint objects, or to discriminate among closely spaced sources of radiation. Nevertheless, as a follow-on instrument to study objects in longer infrared wavelengths, the telescope has achieved a series of remarkable results, including the discovery of the dusty disks of material surrounding the stars Fomalhaut and HR 8799 (see Chapter 6).

Spitzer Explores the First Exoplanet Atmosphere

In 2005, astronomers used the Spitzer Space Telescope to measure Osiris's spectrum from their infrared observations of its star, HD 209458, at two different phases of the planet's orbit. By comparing the infrared spectrum of the star observed during Osiris's transits with observations made half an orbit later, when the star hides the planet, astronomers could rightly ascribe any differences to the planet's existence. Their observations explored the atmosphere of Osiris, the first detected around any exoplanet. In 2001, observations with the Hubble Space Telescope found sodium ions around the planet, and further observations two years later discovered an enormous "exosphere" of hydrogen, carbon, and oxygen that extends to distances several times the planet's diameter. Infrared observations in subsequent years made with both the Spitzer telescope and the Hubble Space Telescope (HST) have identified water vapor, methane, and carbon-dioxide molecules in Osiris's atmosphere.[18] Before we assume that this makes the planet something like Earth, we should recall that its proximity to its star gives Osiris a temperature close to 1,000 K. Temperatures this high make it difficult for a planet to retain its atmosphere: Osiris's hot exosphere presumably represents a continuous outflow from the super-heated exoplanet. Estimates of the rate at which Osiris loses gas imply that over the course of its approximately 5-billion-year lifetime—estimated for its star and assumed to apply to its planet as well—over 5 percent of the planet's mass has been lost through evaporation—the price of remaining so close to its star and one reason why the name Osiris fits the exoplanet.

Osiris reflects only about 4 percent of the incoming light from its star, far less than the 50 percent that Jupiter does. This low percentage suggests that Osiris's upper atmosphere contains compounds that, like the sodium ions already observed directly, reflect light very poorly, though the complete chemical composition in Osiris's atmosphere remains unknown.

Osiris's claims to fame also include its rank as the first exoplanet to be effectively observed directly, once again in the infrared domain with the Spitzer telescope. Unlike the direct images of planets much farther from their stars that can be recorded as points of light, the Spitzer telescope's observations compared measurements of the flux from the star at times when the planet passed directly in front of it with measurements made of the star alone, with Osiris behind it.

MEarth Finds an Earthlike Planet with a Significant Atmosphere

In April 2017, astronomers confirmed that transit observations of the exoplanet GJ 1132b, 39 light years from the solar system, showed definitive evidence of a planetary atmosphere around a "super-Earth," a type of planet similar to, but larger than, our own (described in more detail in Chapter 10). Credit for finding this planet rests with the astronomers who created the MEarth robotic survey, which employs telescopes in Arizona and Chile to search for planets that undergo transits as they orbit low-luminosity red dwarf stars. GJ 1132 easily qualifies, with 26 percent of the sun's diameter, 18 percent of its mass, and $1/1,000$ of the sun's luminosity. Its planet has a diameter about 40 percent larger than Earth's, and an estimated mass roughly 60 percent greater than our planet's.

GJ 1132b, like many planets found by their transits and possessing orbits much smaller than Earth's, takes only 39 hours to orbit its red dwarf at a distance just 1.5 percent of the Earth–sun distance, or 149.5 million kilometers. As a result of its location only 2.25 million kilometers from its low-luminosity star, the planet bakes at temperatures close to 650 K, far hotter than the 300 K that we would regard as balmy. The planet's diameter and mass suggest that its interior could consist largely of silicates with some iron, although the planet could be made largely of water.

In what has now become a fairly routine reaction to Earthbased exoplanet discoveries, a team of astronomers associated with the Max Planck Institute for Astronomy in Germany arranged for the Hubble Space Telescope to observe this star during nine planetary transits, as well as at other times, which allowed them to compare the spectrum of light received with the planet in front of the star with that of the starlight alone. The Hubble observations made in the shorter wavelength-band of infrared radiation found that the planet has a significant atmosphere, made mostly of water vapor, with the possible addition of methane.[19]

KELT-9 b, a Truly Hot Exoplanet

A key search for planetary transits of sunlike stars, managed (mainly) by U.S. astronomers, carries the disarming name of the Kilodegree Extremely Little Telescope, or KELT, which employs two telescopes, located in Arizona and South Africa, to monitor stars over a wide swath of the sky. KELT seeks to find exoplanets sufficiently large for their transits to be observed from the Earth's surface. With lenses measuring only 4.2 centimeters in diameter, which are sometimes swapped for somewhat larger 7.1-centimeter lenses, KELT's twin telescopes essentially provide two telephoto lens systems equipped with CCD detectors to record the stars under observation. The team chose to survey stars intrinsically hotter and more luminous than the sun, and thus significantly more luminous and brighter than the sunlike stars that the Kepler spacecraft observed.

In its survey of several hundred thousand stars, the KELT astronomers have found about two dozen exoplanets. All of these planets qualify as the "hot Jupiters" described in Chapter 4: giant planets that orbit so close to their stars that they produce detectable diminutions in stellar brightness over a short, repetitive cycle. In addition, the KELT team has arranged for other astronomers to measure, so far as possible, the changes in radial velocities induced by the exoplanets' gravitational forces.

On the one hand, this task becomes easier in view of the large velocity changes induced by the orbital motion of a massive, nearby planet, so that the measurements do not require extremely large telescopes and complex spectrographic equipment. Instead of the 1-meter-per-second effects that now lie near the frontier of radial-velocity measurements (see Chapter 4), hot Jupiters cause velocity changes 50 times greater, or even more.

On the other hand, measuring even these sizable velocity changes poses serious challenges in view of the properties of the stars themselves. Highly luminous stars typically rotate rapidly and exhibit a lack of features in their spectrum—two characteristics that seriously increase the difficulty of observing and quantifying velocity changes derived from accurate measurements of stellar spectra. The leader of the KELT enterprise, Scott Gaudi of the Ohio State University, gives high credit to David Latham of the Center for Astrophysics at Harvard University for his determination and ability to tease the desired measurements from the spectral observations. This includes a technique called "Doppler tomography," in which astronomers measure the effects of the planet, as it transits the star, on the individual features of a star's spectrum. The growing list of KELT stars pays tribute to the work of Latham and many others, whose studies have added a new class of stars to those known to possess planets. Gaudi likes to describe the other approaches, which seek to measure ever-smaller changes in radial velocities, as "a race to the bottom."[20]

In 2017, the KELT team announced that one of the hot Jupiters on their short list qualifies as one of the hottest exoplanets now known, with a surface temperature greater than those at which one might naively think that a planet could exist. The star KELT-9, also known as HD 195689, has nearly twice the sun's surface temperature and 50 times its luminosity. KELT-9 b orbits this impressively energetic star at $1/30$ of the Earth–sun distance every 1.5 days. Continuously bathed in vast amounts of visible light and ultraviolet radiation, KELT-9 b maintains a surface temperature just above 4,000 K—hotter than the surfaces of most of the stars in our galaxy. In the gallery of exoplanets, only Kepler-70 b, described in Chapter 8, has a higher surface temperature.

Both its high temperature and the streams of stellar ultraviolet radiation tend to evaporate KELT-9b's gaseous envelope and reduce its mass, but there it is: an exoplanet with 1.9 times Jupiter's radius and 2.88 times its mass. Calculations suggest that the total lifetime of KELT-9 b cannot surpass 600 million years, which makes good sense in view of the fact that a star with KELT-9's luminosity has a lifetime of only about 300 million years.[21] A rival of KELT-9 b for the hottest exoplanet was found by the KELT team's analog across the ocean, the United Kingdom's SuperWASP transit survey program, which employs twin robotic installations, each mounting eight 200-mm lenses, in South Africa and the Canary Islands. This Wide-Angle Search for Planets (WASP), now improved to a Super stage, has been searching for transiting exoplanets around sunlike stars since 2005 and has had more than 100 confirmed successes. All of these planets qualify as hot Jupiters, with orbital periods from 19 hours to 10 days. Unlike the other stars on WASP's exoplanet list, WASP-33 has a mass, luminosity, and surface temperature significantly greater than the sun's, so it would therefore qualify nicely for KELT selection, had the WASP astronomers not already succeeded in finding its planet in 2010.

Located in the constellation Andromeda at 378 light years from the solar system, WASP-33 has 1.5 times the sun's mass and 10 times its luminosity. Orbiting this star every 29 hours, and at a distance just 2.5 percent of the Earth–sun distance, WASP-33 b maintains an average surface temperature of 3,500 K. This giant planet has more than twice Jupiter's mass, sufficient to allow it to retain the bulk of its mass despite its high temperature. By comparing the Hubble Space Telescope's spectroscopic observations of the star during the planetary transits and at other times, astronomers have determined that WASP-33b's outer layers contain both water vapor and titanium oxide.[22]

LHS 1140 b: A Dense, Earthlike Planet with Temperatures Favorable to Life

In 2017, astronomers discovered their current best example of an Earth-like planet within what they call a star's habitable zone (see Chapter 12) by combining transit and radial-velocity observations. Found by the MEarth transit survey, and further analyzed with the High Accuracy Radial Velocity Planet Searcher (HARPS) instrument at the 3.6-meter telescope at the La Silla Observatory in Chile, LHS 1140 b orbits an M star 41 light years away every 25 days, at only 0.088 AU. However, since LHS 1140 has a luminosity only $\frac{1}{50,000}$ of the sun's, its planet receives less starlight energy per second than the Earth does.

Transit observations of LHS 1140 b showed a decrease of half a percent in the brightness of the star, whose diameter equals

19 percent of the sun's. The amount of this dip demonstrates that the exoplanet has a diameter about 43 percent larger, and a volume approximately 2.9 times greater, than Earth's. But the HARPS radial-velocity measurements indicate a planetary mass about 6.7 times greater than Earth's, implying that LHS 1140 b has an average density 2.5 times Earth's 5.51 grams per cubic centimeter! This startlingly large density emphasizes the importance of the words "about" and "approximately"—the always-appropriate scientific recognition that all measurements come with uncertainties. In this case, an uncertainty of 25 percent in the planet's mass implies that the planet's mass might be only 5 times our own planet's, which would give LHS 1140 b an average density about 1.75 times the Earth's. An uncertainty of 7 percent in the planet's diameter could reduce this ratio a bit further. For now, we may conclude that the planet almost certainly has a density significantly greater than Earth's, which seems possible for a planet more massive than ours and therefore capable of squeezing its constituents—dominated probably by iron and closely related minerals—more tightly than our planet does.[23]

Transit Timing Variations Allow Measurement of Exoplanets' Masses

As astronomers have progressively honed their abilities to analyze the data from their observations of transiting exoplanets, they have developed an impressive pathway to deriving the masses of planets that belong to a multiplanet system, which depends on the subtle gravitational effects that the planets induce in their neighbors' orbits. This technique depends on extremely accurate timing of the transits of each planet. If the planets did not affect each other gravitationally, these transits would occur with clockwork-like regularity, reflecting each planet's orbit around the central star. Because these planets do exert gravitational forces on each other, their transit times depend on the locations of the other planets, and they sometimes occur a bit earlier, sometimes a bit later, than the simplified model would imply. The transit-timing variations allow astronomers to estimate the planets' masses—not through their gravitational force on their *stars*, as occurs in radial-velocity searches, but through the forces that the planets exert on each other.

The most stunning examples of astronomers' success in applying this technique to determining the masses of the objects in a multiplanet situation appear in two complex planetary systems: one with six planets and another with seven, one around a sunlike star and the other around a particularly cool M dwarf.

The Six-Planet System of Kepler-11

For a few years, the six planets around Kepler-11 found in 2010 held the record for the largest number of exoplanets known to orbit a star. Three other systems, each with seven known planets (see Chapter 7), have overtaken the Kepler-11 planets, which continue, however, to be the best studied, and in some ways the most interesting, of all known planetary systems that orbit a sunlike star.[24]

Two thousand light years from the solar system, Kepler-11 matches our sun almost perfectly in luminosity and surface temperature, though astronomers estimate its age at 8 billion years, significantly older than the sun's 4.6 billion years. The basic properties of the star's six known planets appear in Table 1. The planets complete their orbits in times that range from 10 to 118 days, orbiting at distances that range from 9 to 47 percent of the Earth–sun distance.

Except for Kepler-11 g, the star's outermost planet, the orbits of Kepler-11's planets would easily fit inside the orbit of Mercury.

Table 1

PROPERTIES OF KEPLER-11'S SIX PLANETS

Planet	Distance from star (AU)	Distance from star (millions of km)	Mass (Earth=1) with probable errors*	Diameter (Earth=1)	Orbital period (days)
b	0.091	13.6	1.9 (+1.4, −1.0)	1.80	10.3
c	0.107	16.0	2.9 (+2.9, −1.6)	2.87	13.0
d	0.155	23.2	7.3 (+0.8, −1.5)	3.12	22.7
e	0.195	29.2	8.0 (+1.5, −2.1)	4.19	32.0
f	0.250	37.4	2.0 (+0.8, −0.9)	2.49	46.7
g	0.466	69.7	< 25	3.33	118.4

*The plus or minus indications for the probable errors of the planets' masses indicate that these errors have different values in the plus or minus directions. Data from Jack Lissauer et al., "All Six Planets Known to Orbit Kepler-11 Have Low Densities," *Astrophysical Journal 770* (2013): 131.

All six planets lie so close to their star that they bake at temperatures well above the boiling point of water.[25]

The fact that all of Kepler-11's six planets undergo transits shows that their orbital planes align to within one degree—an alignment closer to perfection than the orbits of the sun's four inner planets, which align to within about six degrees (Mercury deviates the most). The highly accurate timing of Kepler-11's multiple transits (even the most distant planet makes more than four transits each year) allowed the Kepler scientists to derive the masses of each of them. The innermost five planets have masses from 1.7 to 7.2 times the Earth's mass, with impressively large estimated errors for all but the two most massive. The outermost planet's mass can be established only as less than about 20 Earth masses. Broadly speaking and in view of these uncertainties, we can estimate that all six planets appear to have average densities of matter lower than Saturn's, the most rarefied of the sun's planets. Even with the estimated errors, the planets' average densities exclude the possibility that these planets consist mainly of rock, whose average density typically equals or exceeds 2.5 times that of water. The planets qualify as close-in sub-Neptunes or giant planets; their surface properties and internal structure remain largely conjectural.

Just as we cannot determine the planets' overall appearances, we likewise remain in the dark about their origins. On the one hand, the tight alignment of their orbits implies that they materialized, in agreement with what the basic model described in Chapter 11 implies, from a rotating cloud of gas and dust that surrounded a star as it formed, 8 billion years ago. On the other, we remain ignorant about whether the six planets originated in approximately their present distances, or, like the numerous giant planets close to sunlike stars, assembled themselves at significantly greater distances and migrated afterward into their current small orbits.

We should point out that the Kepler-11 system may have more planets than the six currently known to us. As we have already noted, the Kepler spacecraft's observations cannot find transiting planets whose orbital periods exceed a couple of years. The dip in starlight from a single transit could arise from a number of other causes, and at best it can raise suspicion and draw attention to a particular case. Use of the transit method to find planets at distances from their stars greater than the distance from the sun to Mars must await a new spacecraft, one capable of observing not for four years but for two or three times that interval.

Three Systems with Seven or More Exoplanets

In the interest of providing completeness in describing known multiple-planet systems, we should take a quick look at the three other systems known to possess seven (or even eight) planets each: Kepler-90, HD 10180, and HR 8832.

Kepler-90, a star somewhat more massive and more luminous than the sun though less than half as old, lies about 2,500 light years away in the direction of the constellation Draco. Its innermost six planets bear more than a passing resemblance to Kepler-11's innermost five: They orbit at distances from 7 to 48 percent of the Earth–sun distance, have diameters that equal 1.2 or 1.3 times Earth's for the three inner ones and 2.7–2.9 times Earth's for the next three, and show a tight orbital alignment. No good measurements of the planets' masses exist, however. The two outermost planets have diameters 8.1 and 11.3 times the Earth's, distances from Kepler-90 that equal about 70 percent and 101 percent (!) of the Earth–sun distance (give or take 10 percent), and an upper limit on their estimated masses—the best that we can now provide—about equal to the mass of Jupiter. Thus the Kepler-90 system's known planets include three potentially rocky planets that bake at impressively high temperatures close to their star, three other exoplanets at one-third to one-half of the

Earth–sun distance, and two giant planets that orbit at distances similar to the Earth–sun distance, thus forming a system vaguely akin to the solar system's seven inner planets, but enormously reduced in orbital sizes.[26] In late 2017, the discovery of Kepler-90's eighth planet (third rock from its sun) made this the first system to boast a number of known planets equal to those of our solar system. We may also note that our knowledge of the eighth planet's existence emerged from a "neural network" analysis of the transit data, in which artificial-intelligence techniques that mimic the organizational schema of the human brain searched for signals too faint to be found by conventional techniques. Astronomers naturally plan to extend these analyses to the data for all 150,000 Kepler stars.[27]

HD 10180, another sunlike star 127 light years from us, has seven known planets that were found by radial-velocity measurements with the HARPS spectrographic system. When astronomers lack transit observations, they can derive only minimum values for the planets' masses. These minima range from 1.3 to 66 times the Earth's, with all but the least massive possessing at least 12 times the Earth's, so six of the planets likely have masses between Neptune's 17 Earth masses and Saturn's 95.[28] Giants though they may be, these six planets orbit their star at modest distances, ranging from 0.064 to 3.38 AU, so the system resembles what would exist if we converted all of the sun's inner six planets into giants and shrank their distances from the sun by a factor of 3 or more.

HR 8832's planets, like those of HD 10180, do not transit their star and owe their discovery to the HARPS spectrographic system. Only 21 light years from the solar system, HR 8832, also known as HD 219134, has 79 percent of the sun's mass and 28 percent of its luminosity. The inner four of its six known planets have minimum masses from 4.4 to 16 times the Earth's and orbital distances from 0.039 to 0.147 AU; the outer two have undetermined masses and orbit at 0.375 and 3.11 AU.[29]

The planetary systems around HD 10180 and HR 8832 provide a reminder that without transit observations, the radial-velocity approach can currently reveal only relatively massive exoplanets that orbit comparatively close to their stars. In contrast, the advantages offered by transit observations led to arguably the greatest exoplanet news of 2017: the existence of seven Earth-sized exoplanets in orbit around an M star.

The Ultracool Dwarf with Seven Planets

A collaborative group of exoplanet observers engaged in studying stars at the extreme low end of stellar masses, sizes, and luminosities announced in early 2017 that they had found seven transiting planets in orbit around what we might well call an extreme dwarf. This underluminous star carries the designation TRAPPIST-1, a carefully crafted astronomical acronym that stands for the TRAnsiting Planets and PlanetesImals Small Telescope-South (locating the "I" required some work on the author's part). Created and operated by the University of Liège in Belgium and the Geneva Observatory in Switzerland, the TRAPPIST collaboration employs telescopes in Chile and Morocco; the telescopes' mirrors, just 60 centimeters in diameter, rank among the smallest that astronomers now use for serious research.[30]

Like the Trappist monks of the Cistercian Order, who follow their rules in ongoing silence, the two telescopes continually (and, let us say, silently) monitor the brightnesses of about 60 of the closest faint stars, including the impressively numerous (though unimpressive in their luminosity), ultracool M stars. Billions of years of star formation have strewn these faint stars profusely through our region of the Milky Way, and presumably throughout our galaxy and, we may assume, throughout other galaxies as well, where they remain too faint to be detected. Happily, a search for exoplanets around M stars offers a bonus in comparison to similar searches around sunlike stars. The stars' small sizes ensure that any planetary transit will cause a much more noticeable decrease in brightness than a similar planet's passage across the face of a star like the sun would.

In 2016, the TRAPPIST astronomers studying nearby cool and dim stars with their Chilean telescope found that one of them, 40 light years from the solar system and designated (until astronomers bestowed a better name) as 2MASS J23062928–0502285, showed decreases in its brightness that recurred at regular intervals, the now-familiar sign of transiting objects.[31] Further observations with one of the European Southern Observatory's 8.2-meter Very Large Telescopes established the existence of three transiting planets.[32]

Even for an M star, 2MASS J23062928-0502285, now known as TRAPPIST-1, has a notably small mass and an extremely low luminosity, $1/2{,}000$ of our star's.[33] If this star had a tendency to boast, it might stress that it has four times the luminosity of the least luminous M star so far measured. Its mass, estimated to equal 0.0802 ± 0.0073 times the sun's, or about 84 ± 8 times Jupiter's, puts TRAPPIST-1 close to the dividing line between red dwarf stars and brown dwarfs described in Chapter 6.[34] Current models of these two types of objects set the boundary at approximately 0.082 solar masses, depending on the details of the star's composition and the models used to calculate nuclear fusion in stellar cores.[35] If the Earth orbited TRAPPIST-1, we would receive as much energy per second as we do now from the sun only if we placed the Earth into an orbit 0.023 AU from the star.

Three Planets Turn Out to Be Seven

The discoveries of three close-in planets orbiting an extreme low-luminosity M star led the astronomers who direct the spaceborne Spitzer telescope to study TRAPPIST-1 more closely. This produced startling results: The number of planets rose from three to seven. The outermost of the original three planets turned out to be not one but three planets, whose transits had temporarily almost coincided, along with two somewhat more distant planets that also came to light, or, more precisely, blocked a bit of it. As shown in Table 2, the planets have orbital periods measured in days (just 1.5 days for the innermost and 19 days for the outermost), which imply that their distances from their star, specified as percentages of the Earth–sun distance, equal 1.1, 1.5, 2.1, 2.8, 3.7, 4.5, and 5.9. In more familiar units, these distances vary from 1.7 to 8.9 million kilometers.

As always, the amounts by which the planets' transits decrease the star's brightness provide us with the planets' diameters, which range from 77 to 113 percent of the Earth's. Their masses also (approximately) resemble Earth's, ranging from 41 to 138 percent of our planet's. Like the first planets found with radial-velocity observations, these planets orbit extremely close to their star. By publishing their results in the scientific journal *Nature,* the astronomers and their seven planets provided the world with a brief but welcome diversion from its usual cares: a miniature solar system around a cool dwarf star.

Unlike the close-in planets that orbit sunlike stars, the even-closer-in planets of TRAPPIST-1 do not bake or fry at superheated temperatures. Because the star generates so little energy through nuclear fusion in its core, even the closest of its planets receives only a few times the energy per second that the Earth does. The TRAPPIST-1 planets opened astronomers' eyes a bit wider to the possibility that we may find Earthlike conditions not on exoplanets that orbit sunlike stars, but rather on planets orbiting at a tiny fraction of the Earth's distance from the sun around extremely underluminous stars.

We may note that planet d, orbiting at about 0.021 AU from its star, receives almost the same amount of star-born energy per

Table 2

PROPERTIES OF THE SEVEN TRAPPIST-1 PLANETS

Planet	Distance from star (AU)	Distance from star (millions of km)	Mass (Earth=1) with probable errors	Diameter (Earth=1)	Orbital period (days)	Estimated surface temperature (K)
b	0.0111	1.661	0.85 ± 0.72	1.086	1.5109	400
c	0.0152	2.274	1.38 ± 0.61	1.056	2.4218	342
d	0.02144	3.207	0.41 ± 0.27	0.772	4.0496	288
e	0.02817	4.214	0.62 ± 0.58	0.918	6.0962	251
f	0.0371	5.550	0.68 ± 0.18	1.045	9.2067	219
g	0.0451	6.747	1.34 ± 0.88	1.127	12.3529	199
h	0.059	8.83	undetermined	0.752	18.767	173

second as the Earth does. For the time being, we may feel free to speculate that TRAPPIST-1d and its somewhat cooler neighbor TRAPPIST-1e possess pools, lakes, and seas, since they may have the proper temperatures to maintain liquid water on their surfaces. All that remains would be to check whether this hypothesis can receive verification from future investigations. Spectroscopic observations to be made by the new James Webb Space Telescope, once it becomes operational, may provide just such an opportunity (see Chapter 12).

Orbital Resonances and Planetary Masses

The orbital periods established for these planets exhibit what astronomers call orbital resonance, a phrase reminiscent of the harmonic resonance of sound waves. Orbital resonance describes a situation in which two orbiting objects have periods with a simple ratio, such as 2:1 or 3:2. This does not occur by accident; instead, the gravitational interaction between the orbiting objects gradually changes their orbits and establishes their resonance ratio. In the solar system, for example, among Jupiter's three innermost large moons, the innermost, Io, orbits twice every time that the next moon, Europa, orbits once, and Europa orbits twice every time that the third-innermost moon, Ganymede, orbits, once. This means that the orbital periods of each neighboring pair of moons have a resonance ratio of 2:1. The orbital periods of the six innermost planets orbiting TRAPPIST-1 have five slightly more complex orbital resonances, with ratios close to 8:5, 5:3, 3:2, 3:2, and 4:3.

As these six closely spaced planets tug on one another, they generally maintain their orbital periods, but these periods undergo small changes that astronomers can detect by carefully timing their transits and thus deducing the planets' masses, as we have already described. The planets' estimated surface temperatures, which depend directly on their distances from their star (neglecting the effects of any atmospheres that they may possess) vary from 400 K for the innermost planet to 173 K for the outermost, and for two of them, the second- and third-innermost, these temperatures (288 and 251 K, respectively) lie within the range (273 to 373 K, or 0 to 100 Celsius) at which water remains liquid under one atmosphere of pressure. These factors naturally provoke speculation that living organisms might now be found on at least two of the seven planets, prime targets for close inspection by the spacecraft of our future (see Chapter 14).

The downside of inclining toward the conclusion that TRAPPIST-1 offers planets favorable to life centers on the planets' impressively short orbital periods and small orbital distances; all of them have become tidally locked to their star, so that they maintain one perpetually lit and one perpetually dark hemisphere. (The astronomer Courtney Dressing has remarked that if civilizations exist on planets like these, the astronomers could live happily on one half, with everyone else on the other, beach-y side.[36]) Also, the K2 spacecraft observed a flare from the star equal in energy output to the terrestrial Carrington event described in Chapter 12, which could pose serious danger to any of these civilizations. Because the TRAPPIST-1 observations extend over only a few years, in comparison with the centuries on Earth that recorded a single comparable event, its flare confirms astronomers' knowledge that M stars often undergo extremely energetic outbursts that would threaten danger to life on any planets close to these stars. The stars can also emit streams of ultraviolet radiation that will impinge strongly on planets orbiting at small distances, threatening forms of life similar to those on Earth. Close to four billion years ago, when the sun generated much more ultraviolet radiation than it does now, the earliest, simplest, and most fragile forms of life may have evolved deep underwater, where they were protected by tens of meters of ultraviolet-absorbing liquid. As Carl Sagan wrote as long ago as 1961, "it is likely that the first self-replicating poly-nucleotides [on Earth] developed in the oceans, and were benthic rather than pelagic."[37]

A System Ripe for Panspermia

The mutual proximity of the seven TRAPPIST-1 planets heightens speculation that if life had ever originated on one of these worlds, it might well have managed to spread to one or more of its neigh-bors. Although the term "panspermia," introduced by the ancient Greek philosopher Anaxagoras, may seem a bit overblown in this context because its etymology implies that life's seeds permeate the universe, it has come into use to describe transfers of living organ-isms from one world to another. Applied to the solar system in some detail in 1903 by the Swedish chemist Svante Arrhenius, the panspermia hypothesis deals with the possibilities that life might have come to Earth from other worlds, or that Earth life could establish itself on the neighboring planets Venus and Mars.[38] This remains a significant and viable proposition, to be well tested when and if we find living organisms on Mars, or anywhere else in the solar system, and compare their basic structure and biochemistry with those of their terrestrial counterparts.

In the solar system, neighboring worlds occasionally approach one another to within less than 50 million kilometers, hardly a close encounter for the purpose of transferring life from world to world. We may imagine, however, that microscopic organisms en-capsulated in meteoroids blasted from a planet's surface might survive for millions of years, eventually to reach another world and seed it with life. In the TRAPPIST-1 system, planets approach one another to within one million kilometers, or even less, increasing the probability of material passing from world to world.

Improvements in Exoplanet Transit Observations

The Kepler spacecraft's trailblazing success has made it clear that transit searches of wider areas of the sky, and of fainter stars, have the potential to find thousands upon thousands of additional exoplanets. Transiting planets will always reveal only a small fraction of the number of similar planets that orbit similar stars at similar distances but happen to move in orbits that don't carry them across our lines of sight to their stars. Nevertheless, the sheer numbers and variety of the planets soon to be found by the transit method should continue to allow astronomers to refine and to round out their statistical understanding of exoplanets in our sun's neighborhood.

The near future should bring into operation four separate detection systems that rely on the transit method. Two of these, created at far less expense than the other two, will remain on the ground, while the two spaceborne instruments, TESS (the Transiting Exoplanet Survey Satellite) and CHEOPS (the CHaracterising ExOPlanet Satellite), launched in April 2018 and ready for launch in 2019, respectively, will employ the clarity of space to probe for small transiting exoplanets.

TESS will make the first all-sky search for Earth-sized transiting planets that may orbit around half a million of the brightest stars in our galactic neighborhood, including a thousand M stars, by measuring the stars' brightnesses every few minutes. The transiting planets' sizes and masses will allow astronomers to create an accurate census of Earthlike planets in our corner of the Milky Way.[39]

CHEOPS, the other spaceborne mission to study exoplanets by their transits, will build on previous results by studying only bright stars for which radial-velocity observations have already found orbiting planets. Created by the Swiss Space Office and the European Space Agency, CHEOPS will make extremely accurate brightness measurements of this comparatively limited set of target stars as it seeks to characterize transiting planets—most directly by measuring their sizes—that have masses from one to six times the Earth's. CHEOPS will therefore obtain an accurate picture of the frequency with which these types of exoplanets appear in our sun's neighborhood.[40]

Back on Earth, the Next Generation Transit Survey (NGTS), constructed by a consortium of institutions in Germany, the United Kingdom, and Switzerland at ESO's Paranal Observatory in northern Chile, swung its 12 robotic telescopes into action in 2015. Each of these telescopes has a mirror only 20 centimeters across, which allows the telescope array to survey an area of 96 square degrees on the sky, about 500 times the area of the full moon. Within its field of view, the NGTS can monitor the brightnesses of several thousand stars. Because the clear air above the observatory introduces an uncertainty in the measurement of stellar brightnesses by about one part in a thousand, the NGTS can detect only planets significantly larger than Earth, in orbit around stars brighter than those observed by Kepler.[41]

The other ground-based planet-hunting array of telescopes aims to survey the entire sky, not the 96 square degrees that amount to a bit less than 1/400 of the total area, using telescopes in Chile and California. Project Evryscope, funded by the National Science Foundation, endowed with a name that its leaders claim to mean "wide-seer," sets 27 telescopes on a single mount to secure 2-minute exposures of the entire sky throughout the night. The project's southern component, located a few hundred kilometers south of the NGTS at the Cerro Tololo Interamerican Observatory, began operation in 2015; its complementary northern installation began construction in 2017 at the Mount Laguna Observatory in southern California. Once both locations become fully operational, Evryscope will effectively create a nighttime movie of everything in the sky, night after night, for as long as its detectors—780 million pixels strong—remain operational.[42]

10

⬩

WHAT HAVE WE LEARNED?

ach of the big three of exoplanet search techniques—
radial-velocity measurements, transit observations, and (less
successful in sheer numbers) gravitational microlensing—
bring important advantages and disadvantages to the quest for
other worlds. Improvements in radial-velocity techniques, together with the flood of discoveries from transits observed by
the Kepler spacecraft, have demonstrated that although hot Jupiters naturally turned out to be the first known exoplanets, they in
fact amount to less than half a percent of all exoplanets (see
Figure 9).

Biases Created by Different Techniques

The hot Jupiters remind us that as we attempt to assess the true
distribution of different types of planets in our neighborhood, we
should bear in mind the biases that each method of searching for
exoplanets embodies. Radial-velocity studies have proven the most

successful in finding massive planets close to their stars, though they can also unveil massive planets at large distances from them, or lower-mass planets near their stars. In addition, radial-velocity observations provide a planet's mass, multiplied by an unknown factor (less than or equal to 1) that depends on the system's orbital alignment. Kepler could find transiting planets at least as large as Earth, in orbits up to the size of Earth's, and furnished a direct measurement of an exoplanet's diameter, allowing derivation of its mass based on an estimate of the planet's composition. Radial-velocity studies of a star with a transiting planet eliminate this uncertainty, providing astronomers with the exoplanet's actual size and mass. Both the radial-velocity and transit techniques reveal a planet's orbital period and thus the size of its orbit. Microlensing claims a much shorter discovery list of planets, but it contributes significantly to completing the exoplanet census through its ability to find planets that orbit comparatively far from their stars, without the transit method's requirement of a nearly exact alignment of a planet's orbit with our line of sight.

The finest precision in measuring velocities along our lines of sight that astronomers can make currently, somewhat better than 1 meter per second, allows them to find a planet with a mass like Neptune's (17 Earth masses) around a sunlike star only if the planet orbits about as close to that star as the Earth does to the sun. Future instruments, some of which are described in Chapter 13, will improve the precision of these measurements and allow astronomers to find less massive planets, and planets at greater distances from their stars, than they can now.

Although microlensing surveys carry a bias against detecting planets with masses as small as the Earth's, they have shown that about one-third of all stars have planets with masses very roughly comparable to Saturn's, and that among these stars, lower-mass, Neptune-like planets are more common than higher-mass, Jupiter-like planets.

Super-Earths and Sub-Neptunes

Kepler's observations have added to its most basic finding—that the majority of sunlike stars in the Milky Way galaxy have at least one planet—to include the facts that planets found by their transits typically have masses 1 to 10 times the Earth's and move in orbits with sizes from 0.02 to 1 AU.[1] Many planets have masses much larger than these (because they are the easiest to detect, their numbers come with some bias), and some planets have orbits much larger than Earth's, with more discoveries to come as astronomers have more time to observe their stars and the ability to make more accurate radial-velocity measurements.

The most significant recent development in exoplanetary demographics has emerged from more precise determinations of the sizes of the Kepler stars, which in turn have allowed astronomers to derive more accurate measurements of the sizes of their transiting planets. These improvements have led to the realization that the most common sorts of exoplanets—planets that have orbital periods of 100 days or less, which puts them comparatively close to their stars—can be divided into two populations, "super-Earths" and "sub-Neptunes."[2]

This conclusion has been confirmed through the impressive efforts of a team of astronomers who formed the California-Kepler Survey (CKS) in order to determine the diameters of 2,025 stars with Kepler planets in orbit. Led by Benjamin Fulton of Caltech and the University of Hawaii, the CKS team inferred the diameters of these stars from spectroscopic observations made with the HIRES spectrometer, developed by Stephen Vogt and his collaborators, at one of the twin Keck telescopes.[3]

The CKS results show that the great majority of Kepler planets with orbital periods of less than 100 days—in other words, planets that orbit their stars at distances less than Mercury's distance from the sun—have diameters between 1.2 and 3.1 times Earth's. Collectively, these exoplanets form a class of planets unknown in the solar system; indeed, these planets were not even suspected to exist in such large numbers before Kepler had its way with a 150,000 stars in the Milky Way. Although Kepler found significant numbers of planets with diameters comparable to those of Neptune, Jupiter, or even larger planets, the "basic Kepler planet" in a fairly close-in orbit has a diameter intermediate in size between those of the Earth and Neptune. We should note for completeness that at the smallest range of their sizes, many transiting planets may have passed unnoticed.

In an even more surprising result, one that ranks high on the list of the hot exoplanet news of 2017, the CKS found that the size distribution among the basic Kepler planets with orbital periods less than 100 days presents a clear gap in the size range between 1.5 and 2.0 times the Earth's. As shown in Figure 11, exoplanets with sizes less than 1.5 times Earth's, or between 2 and 3 times Earth's, have proven to be far more common than those with intermediate sizes. The diameters of the Kepler planets concentrate toward one of two values, 1.3 or 2.4 times the Earth's, with approximately equal numbers in each category.

What has created the exoplanets' division between super-Earths and sub-Neptunes? The favored explanation draws on astronomers' models of how planets form, and on planets' ability to retain hydrogen and helium gas. These considerations have led astronomers to propose that the sub-Earths represent the largest of the rocky planets that failed to retain an atmosphere that would have given them significantly larger sizes. In contrast, the sub-Neptunes have atmospheres made mainly of hydrogen and helium gas that do add significantly to the exoplanets' sizes without adding much to their masses. Presumably, rocky planets that grew to sizes less than 1.8 times the Earth's lacked the ability to retain the two lightest elements, whereas rocky planets that grew to sizes

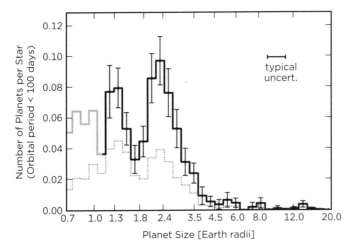

Figure 11 The heavy line in this diagram groups the Kepler-discovered planets with orbital periods less than 100 days into "bins" specified by their sizes. The gap at planetary radii between about 1.5 and 2.0 times the Earth's radius separates the exoplanets known as "super-Earths" from those denoted as "sub-Neptunes." The horizontal line in the upper right shows the uncertainty in determining the sizes of the planets, and the vertical lines denote the uncertainty in numbers. (Plot reproduced from Benjamin J. Fulton, Erik A. Petigura, and Andrew W. Howard et al., "The California-*Kepler* Survey, III. A Gap in the Radius Distribution of Small Planets." *Astronomical Journal* 154, no. 109 [September 2017]: 1–19. Courtesy of Benjamin Fulton)

above this boundary could hold on to hydrogen and helium, achieving sizes greater than twice the Earth's.

The exoplanet expert David Kipping has pointed to an interesting aspect of the newly established super-Earth category: our planet's rank among the largest of these objects. As Kipping puts it:

> The emerging evidence for a divide between rocky planets and gaseous planets at [about]1.5 Earth radii was quite surprising and generally everyone expected this divide to occur at more like 2 Earth radii. That might sound like a small difference, but it corresponds a quite a large difference when translated into

mass. . . . In fact there is compelling evidence that the dividing line may even be closer to 1.2–1.3 Earth radii. What this means is that the Earth may actually represent one the largest rocky planets in the universe; in other words, we may be a Super-Earth all along. This kind of makes sense . . . [because] we would generally expect to live on [one] of the largest [rocky] planets since it supports a larger population.[4]

The Numbers of Multiple-Planet Systems

The CKS examined 2,075 planets confirmed to transit 1,305 Kepler stars, of which many hundred belong to multiple-planet systems, and concentrated on 1,385 planets orbiting 960 stars.[5] Models of how stars and planets form predict that most multi-planet systems should have their planets' orbits in almost the same plane, as occurs in the solar system. (Testing this conclusion, however, would require the ability, which we currently lack, to study planetary systems that fail to follow this prediction.) The models therefore imply that the existence of a single transiting planet should indicate the greater likelihood of success from a detailed search for additional planets around that star.

About 250 of the planetary systems cataloged by the CKS contain two known exoplanets; 84 systems have three; two dozen have four; and the remaining two dozen include five, six, or seven known exoplanets. All of these numbers must be regarded as lower bounds, since smaller planets, and planets significantly more distant from their stars than those detected in transit, may well have escaped detection. This caution about the biases inherent in every technique applies to all methods of searching for exoplanets, though each method best favors the detection of particular types of planets.

Before we turn our attention to the formation of exoplanets, we should note that our sun's planetary system shows a fundamental resemblance to most of the other systems that have been discovered so far, even though it differs markedly in many significant aspects. Unlike the decades preceding the end of the twentieth century, when astronomers could reasonably argue that the existence of the solar system might represent a highly anomalous situation in the Milky Way, today astronomers do not doubt that a multitude of planets bestrew the starry cosmos. Radial-velocity and transit observations imply that most stars have planets, and that multiplanet systems—more difficult to detect than single planets—likewise appear in large numbers. In these two features, our solar system ranks among the multitude.

In broad terms, we can proudly proclaim that at least half of all stars have one or more planets with orbital periods less than 100 days. (Note that the sun technically qualifies only because of the 88-day period of Mercury, a planet far too small to have been discovered by our current basic techniques.) Among the larger planets, with masses more than five times Earth's, smaller planets outnumber the larger ones. When astronomers perform a statistical extrapolation from the available observational data, they find that it is likely that about one-fifth of all sunlike stars possess one or more giant planets in orbits up to 20 AU in size. Among smaller planets, the distribution by sizes forms a plateau, divided into the super-Earths and sub-Neptunes, as shown in Figure 11. The concentration of modestly sized planets with short orbital periods shown in Figure 9 may arise mainly from the restrictions of observational techniques, so that planets in larger orbits may eventually prove to be about as common as those in smaller ones. Finally, giant planets appear preferentially around stars with higher concentrations of elements heavier than hydrogen and helium.

These summary sentences testify to many effortful years, combined with billions of dollars expended on telescopes, spacecraft, and data reduction, not to mention astronomers' salaries. Provided that this support continues and even grows, another generation of readers should have easy access to a full knowledge of the orbs of heaven that Alexander Pope imagined in 1733 in his poem "An Essay on Man":

> Through worlds unnumber'd though the God be known,
> 'Tis ours to trace him only in our own.
> He, who through vast immensity can pierce,
> See worlds on worlds compose one universe,
> Observe how system into system runs,
> What other planets circle other suns,
> What varied being peoples ev'ry star,
> May tell why Heav'n has made us as we are.[6]

11

·

HOW PLANETS FORM
WITH THEIR STARS

Before astronomers knew many facts about exoplanets, they understandably hoped that learning more about our planetary cousins would help to confirm their basic model of how the solar system formed, and that the comparison of key details of the sizes and orbits of exoplanets with those of the sun's family would provide important insights into the formation processes.

As often occurs, scientific progress proceeded along a path somewhat different from earlier expectations. The 500-fold increase in the number of known planets has allowed astronomers to find exoplanets with highly extraordinary—by solar-system standards—sizes, masses, orbits, mutual interactions, and home stars. Although the solar system continues to furnish important and useful information about the general nature of planets, the substantial contrasts between the solar system and the exoplanetary zoo verify the old saying that in reaching conclusions, 1 is a dangerous number. While keeping that warning in mind, let us consider how the solar system began.

The Formation of the Solar System

Although controversy remains about the events that created the sun and its planets, almost no astronomers doubt that the bulk of this evolution from primordial dust and gas began about 4.55 billion years ago; that the process occurred rapidly, in astronomical terms; and that almost all solar-system objects formed concurrently. Dating the start of the solar system draws on the ages of the oldest meteorites found on the Earth's surface; the known decay rates of radioactive minerals contained in those meteorites have established that their ages reach 4.568 billion years. Among larger solar-system objects, we have ages for just two: the Earth and its moon, upon which the oldest rocks have ages of 4.4 billion years for Earth (with very few examples and somewhat controversial age determinations) and 4.44 billion years for the moon. The near coincidence in age of the oldest meteorites and lunar and terrestrial rocks, together with the slightly older ages for the meteorites, adds weight to a model in which all of the objects in the solar system—the planets, their satellites, the asteroids, and the billions of comets and meteoroids that orbit the sun—formed through the coalescence of small objects into larger ones that occurred within an astronomical blink of an eye, probably less than 100 million years, or even less than 10 million years, at a time just over 4.5 billion years ago.[1]

In the generally accepted model of the solar system's formation, an enormous cloud of gas and dust began to contract nearly 4.6 billion years ago. Because the cloud had some initial rotational tendency (so the model assumes), its contraction, following the laws of physics, caused it to flatten and to rotate more rapidly. These effects increased as the cloud contracted further, so that it soon became a flattened pancake of gas and dust, densest at its center, rotating most rapidly in the regions closest to its center and most slowly in its outer extremities.

Within this "protoplanetary disk," subclumps formed the planets and their large moons, as dust grains aggregated to form small particles, which coalesced to make pebbles and rocks, which soon came together to form much larger solid objects. At the center, the clump with the dominant mass continued to contract under its self-gravitational forces. The contraction heated the protosun, most notably at its center. When the protosun's central temperature reached about 10 million K, nuclear fusion began within its core, marking its official transition from a protostar to an actual star.

The heating induced by the protosun, followed by still greater heating from the nuclear-fusing sun, hastened the evaporation of nearly all the gas remaining around the sun's inner clumps, which evolved into the four inner planets and the moon. Beyond the orbit of Mars, a host of solid objects, also deprived of surrounding gas, failed to coalesce to form a planet, probably because of the disturbing influence from the gravitational force of the nearby giant planet Jupiter. These objects became the asteroids that still orbit, for the most part, between Mars and Jupiter.[2]

At greater distances from the sun, we encounter what astronomers call the "snow line" or "ice line": the distance from a star beyond which temperatures fall to the point that at least some of the most common volatile compounds—methane, ammonia, carbon dioxide, carbon monoxide, hydrogen sulfide, and water—can condense into solids. Astronomers' use of the word "ice" includes all of these molecules and their compounds if they achieve solid form. Inside the snow line, astronomers expect that planets will form from the cosmically abundant elements silicon, oxygen, iron, and aluminum, with smaller amounts of calcium, magnesium, and sodium, whose compounds dominate terrestrial rocks. Beyond the snow line, mixtures made of rock and ice, or mainly of ice itself, should provide significant contributions to the mass of any planet, including the cores of giant planets. In accord with this rule, we expect that each of the sun's four giant planets, which orbit at distances that range from 5 to 30 times the Earth–sun distance, first built a solid core from rock or a rock-ice mixture. The gravitational force from their cores allowed these proto-planets to gather and retain a large amount of the gas in their vicinities. Most of the mass of Jupiter, Saturn, Uranus, and Neptune now consists of hydrogen and helium, the two lightest and most abundant of all elements, which compose 99 percent of the sun's mass and a smaller proportion, though still a sizable majority, of the giant planets' masses.

In earlier versions of the formation model, the cores of the giant planets resembled super-sized versions of the inner planets, hidden forever by much greater masses of hydrogen and helium gas surrounding them. Modern investigations, most notably by the Juno spacecraft now in orbit around Jupiter, change this picture somewhat by suggesting that the planet has a "fuzzy core" that could have a radius half as large as Jupiter's, containing large amounts of water, ice, and frozen methane along with rocky minerals similar to Earth's.[3]

Outside the orbit of Neptune we find the most pristine members of the solar system, those least affected by the 4.5 billion years of recent history, continuing to orbit along their frigid paths around the sun. Collectively these objects form the "Kuiper belt," named after the Dutch-American astronomer Gerard Kuiper, who first deduced their existence. Many billion comets, balls of rock and ice (in the astronomical sense), measuring a few or a few dozen kilometers across, orbit in this distant darkness; so, too, do Pluto, Eris, Makemake, Quaoar, and other small rocky worlds, most of which, we may reasonably speculate, remain to be discovered.

This model makes coherent sense, although it leaves (for such is the nature of scientific inquiry) a goodly number of loose ends and unresolved mysteries. Within the disk of gas and dust, calculations show that individual dust particles could indeed stick together quickly and efficiently through electrostatic forces to form increasingly larger dust grains, pebbles, and even rocks. However, the coalescence of these objects into "planetesimals," rocky objects a kilometer or so in size, like the smallest asteroids, would take much longer—too long to fit within the time frame implied by the ages imposed by the model. Because individual rocks are much farther apart than the millions of smaller particles from which they formed, they collide and coalesce much more slowly than the dust grains did. The possibility that the some of the larger dust grains may have begun to form even before the protoplanetary disk began to condense helps to support this model only modestly.

NASA's Jack Lissauer puts things in a nutshell: "The formation of planetesimals remains very controversial."[4] One of the astrophysicists now seeking a solution to the rocks-into-planetesimals problem, Konstantin Batygin, can boast (though he does not) fluency in three languages, since he spent his early childhood in Moscow before his father, an accelerator physicist, took the family to Japan for five years before settling in California, where Batygin won an undergraduate prize for his astrophysics thesis at the University of California, Santa Cruz, and, of course, played in a rock band. In 2016, Batygin, a Caltech PhD and assistant professor of planetary science, worked with Michael Brown in advancing the hypothesis of Planet Nine (see Chapter 3), and he was named one of the 10 most brilliant people of the year by *Popular Science* magazine.[5]

Batygin and others who investigate models of the solar system's formation suggest that the larger dust grains in the protoplanetary disk were slowed by encounters with the disk's gaseous component, causing them to drift inward. As they did so, the grains underwent a "peloton effect," analogous to the lower air resistance felt by the followers in a bicycle race. This allowed the grains in the peloton to congregate to form ever-larger clumps; the process continued until the clumps became planetesimals more than a kilometer across. The individual gravitational forces of those planetesimals then began to play an increasingly significant role.

From then on, the planetesimals grew into more massive objects either through their mutual gravitational attraction, or by attracting all the rocks in their vicinities, or both. As their masses increased, the planetesimals' gravitational forces eventually led to a "runaway accretion" that gathered almost all the solid material into a few objects: the planets and their largest moons. Beyond the snow line, objects with at least 5 or 10 times the Earth's mass would rapidly gain mass from the large amounts of hydrogen and helium gas in the protosolar disk, soon becoming the giant planets that we see today.[6]

Does Batygin's model correspond well to reality? So far, so good, if one follows the peloton to bridge the gap between rocks and planetesimals; the concept that a less massive core soon attracts more material to increase its mass significantly seems entirely likely with or without a peloton. Attempts to explain the production of large objects nevertheless leaves some serious issues. Here's one: How do we explain the fact that the moon contains a mixture of chemical elements and their isotopes that closely resembles, without being identical to, the mix of elements and isotopes in the Earth? The difference may be modest, but it forces astronomers to reject the pre-lunar-exploration postulate that the moon represents a large chunk of the proto-Earth that broke away as our planet formed. The current best hypothesis holds that a Mars-sized object struck the Earth 4.5 billion years ago, smashing itself to bits and mixing some of its material with matter blasted away from the Earth. Thus the composition of the moon that coalesced nearby would approximate our planet's distribution of elements without duplicating it.

Another problem with the basic model centers on Mars, or, more precisely, centers on the fact that Mars has only one-tenth of the Earth's mass. Because the region around Mars's orbit includes a greater volume than that surrounding the Earth's, Mars presumably had a chance to acquire a greater amount of mass in the proto-solar system than Earth did. Why did Mars end up with so little? An analysis made by Harold Levison of the Southwest Research Institute and his colleagues answers this question with the concept that the pebbles and rocks beyond the Earth's orbit would have been moving too rapidly for planetesimals to attract them efficiently—in contrast to the situation more conducive to such captures that prevailed at distances from the sun significantly less than Mars's.[7]

These and other issues connected with the basic accretion model for forming the sun's inner planets and the cores of the giant planets will require further investigation to achieve full resolution. Nonetheless, the model has achieved reasonably good success in explaining how the solar system achieved its present configuration. Certain refinements of the model have the giant planets changing their orbits, and even their order outward from the sun, as the result of their mutual gravitational interactions; strictly speaking, this has little to do with the giant planets' actual formation, but it explains why we find them in their current orbits.

Not all astronomers have joined the chorus in support of a model in which all members of the solar system, or at least the cores of the largest ones, formed through the gradual accretion of smaller objects to create larger ones. A competing model envisions that instabilities within the protoplanetary disk induced sudden collapses, leading to clumps of matter with the mass of a giant planet. The instability models seem to have a better chance to explain the formation of larger planets at comparatively great distances from their stars, where greater amounts of material exist.

In particular, the instability process favors the genesis of the most massive exoplanets—those with masses thousands of times the Earth's mass and 4 to 30 times Jupiter's—that form far from their stars. Future observations with the spaceborne telescopes described in Chapters 10 and 13—TESS, JWST, and Gaia—should help to determine whether the bifurcated model of planet formation describes the host of planets in the Milky Way.

ALMA Observes Protoplanetary Disks

In recent years, excellent direct evidence has emerged that planetary systems do form, even now, within flattened protoplanetary disks similar to the one believed to have given birth to the solar system. Most notably, the marvelous ALMA array, the crowning glory of astronomers around the world who secured its funding, has allowed astronomers to observe examples of this star and planet formation as it occurs. The Atacama Large Millimeter Array (ALMA) observatory in northern Chile consists of 66 dishes that capture millimeter-wavelength radiation, which occupies the region between the radio and infrared portions of the electromagnetic spectrum. Millimeter radiation, typically generated by extremely cool sources, contains a host of useful information. Astronomical reports bring comparatively little news from this spectral domain because our atmosphere absorbs millimeter radiation almost completely, preventing it from penetrating to altitudes much below 5,000 meters.

ALMA was built in response to these facts of life. Completed in 2013 at a cost of $1.4 billion, constructed by a consortium that includes the European Southern Observatory, Japan, the United States, Canada, South Korea, and Taiwan, ALMA now ranks as the most expensive ground-based telescope ever constructed (see Chapter 13 for plans to surpass this outlay). Under Chile's dry skies at an altitude of 5,059 meters, higher than any mountain in the Alps or the lower 48 states, ALMA's 66 dishes (54 of them measure 12 meters across, 12 of them are 7 meters in diameter) profit from the almost total lack of atmospheric water vapor that would prevent observation of the millimeter-wave radiation that dust grains emit at temperatures 10 or 20 degrees above absolute zero. Construction of this billion-euro project presented serious operational challenges, starting with the difficulty of keeping one's head clear at 5,000 meters. Today the site's administration building allows astronomers and technicians to breathe more easily, not by artificially increasing the atmospheric pressure, which would require airlocks and much added expense, but by the more ingenious approach of employing special filters that enrich the fraction of oxygen in the building's air.[8]

Figure 12 This nighttime view of part of the ALMA array shows the southern skies' chief fuzzy ornaments, the two satellites of our galaxy called the Magellanic Clouds, beyond the closest dish. (Courtesy of the European Southern Observatory / C. Malin [cc BY 4.0])

ALMA's observational triumphs include a plot of millimeter-wave emission in the object that astronomers call TW Hydrae, in which dust-rich gas surrounds a star only a few million years old. ALMA's map of the distribution of the dust within this "proto-planetary disk" revealed notable gaps suggesting a "clearing-out" process similar to what happened—so we think—within the sun's own protoplanetary disk.

At 195 light years from the solar system, TW Hydrae represents our closest current protoplanetary neighbor. The gaps in the outer regions of ALMA's map lie at distances from the central star of about 3 and 6 billion kilometers, close to the sun–Uranus distance and to ⅘ of the sun–Neptune distance, respectively. In addition, a gap in the inner regions of the disk has a distance close to the sun–Earth distance.[8]

Forming Exoplanets

The discovery of several protoplanetary disks, of which TW Hydrae offers the best example, helps to verify the model for the initial stages of all planetary systems. They certainly support astronomers' conviction that almost all exoplanets formed along with their stars, just as the sun's planets did, slightly more than 4.5 billion years ago in the case of the solar system.

We might decide that everything beyond this stage—planets, moons, and smaller objects form with different sizes, masses, densities, orbits, and so forth—is detail that astronomy graduate students and their supervisors can carefully examine. Or we might rather conclude that the similarities and differences in the world of exoplanets cry out for explanations, which have been supplied only in their initial and controversial aspects. The passage of time should allow both currents of thought to converge toward a happy outcome. An argument can be made that we should refrain from too rapidly adopting a need for explaining a particular phenomenon, when much larger phenomena may soon appear on the scene.

Consider, for instance, the great problem posed by some of the first exoplanets to be discovered, those with masses greater than Jupiter's but distances from their stars less than about 0.1 AU. How could the model based on our solar system explain the existence of these "hot Jupiters" so close to their stars? From where would their great mantles of gas have come? How could such planets avoid evaporation under the intense heat from their stars?

Astronomers found reasonable explanations for most of the problems posed by hot Jupiters, whose numbers now approach 100. Equally important to note, however, are the facts that (a) hot Jupiters amount to a small minority of all known exoplanets, and (b) some exoplanets much farther from their stars raise equally difficult, and in some cases more significant, problems of explanation.

Hot Jupiters and Planetary Migration

Astronomers knew that the radial-velocity technique would bias their findings toward the most massive planets that orbit closest to their stars. Nevertheless, they received a deep and genuine shock when the first nonpulsar exoplanets turned out to possess not only masses comparable to, or larger than Jupiter's, but also startlingly small orbits.

What could explain a giant planet that orbits its star not at 5.2 AU, as Jupiter does, or at 0.37 AU, as Mercury does, but at only 0.05 AU? As Isidor Rabi once asked in a different context, who ordered that? The planetary organization embodied in astronomers' favored models seemed to collapse in the fall of 1995, first with a single exoplanet, then with five more, and before long with dozens after that.

On one key point, all astronomers involved in creating solar-system models on their computers and in their capacious minds have found agreement: Giant planets orbiting at distances much less than the Earth–sun distance can hardly have formed so close to their stars, unless our current understanding of what occurs to make a planet should turn out to be largely misguided. Instead, hot Jupiters must have accomplished the bulk, if not the totality, of their formation much farther from their stars than their current orbits. Some process or processes must have moved these giants from the much larger orbits in which they originally formed into orbits that span only $\frac{1}{10}$, $\frac{1}{100}$, or an even smaller fraction of their original sizes.

Astronomers who had worked so long on models of how stars and planets form provided answers that sprang from the concepts some had already conceived. Today, confronting the grand array of planetary sizes, masses, and orbital parameters, the astronomical world can boast of three different ways to explain how some giant planets came to find themselves in close proximity to their starry overlords. All three explanations depend on the basic force of the cosmos at large distance scales: gravity. If you want to shrink the orbit of a super-Jupiter by hundreds of millions of kilometers, you have better rely on gravity, for nothing else seems remotely capable of doing the job. Each of the three different avenues to achieve this effect proposes a third object, or quasi-object, in addition to the star and its planet. Those objects are (1) a binary star system, in which a second, smaller star exerts gravitational force on a planet; (2) one or more additional planets, whose gravitational forces likewise induce a neighboring planet to move; or (3) a disk of material, left over from the formation of the star and its planets, that orbits the star inside the original orbit of the planet in question.

Let us skip over the details of possibilities (1) and (2), which are known as "dynamical migration." These explanations offer the simplicity of imagining one object attracting another by gravitational forces and making it move, which seems immediately reasonable. In contrast, possibility (3) raises the more complex possibility of "tidal migration," or moving a planet through the exoplanet's gravitational interaction with protoplanetary material that will never form a planet. Imagine a giant planet that has formed in a roughly Jupiter-sized orbit, at a time when significant amounts of material remain in the protoplanetary disk, both inside and outside the young planet's orbit, so that the situation resembles that shown in ALMA's map of the dusty material around TW Hydrae. Although the diffuse gas and dust in this disk exert relatively modest gravitational forces on the planet, nevertheless, in the fullness of time (hundreds of thousands of years), their effects can make the planet move into a different orbit.

Like the velocity of the planet, the orbital speeds of the material in different parts of the disk will depend on the distance from the star. Lesser distances inevitably require greater speeds to maintain the balance between momentum and gravity that keeps each object or particle in orbit. Because of the differences in orbital velocities, the gravitational forces exerted on the planet by the portions of the disk closer to the star tend to make the planet move more rapidly; in contrast, the material outside the planet's orbit tends to slow the planet. Although this may contradict intuition, a tendency to speed up will make the planet move inward, whereas an influence that tends to slow the planet will cause it to move outward. These differences can induce planetary migration, but only if a significant difference arises between the amounts of nearby material *inside* and *outside* the planetary orbit.[9]

These differences can arise as the diffuse material in the protoplanetary disk undergoes friction within its various components. The friction tends to make the material slowly spiral toward the star (see the discussion of planetesimal formation earlier in this chapter). This spiraling inward tends to open a gap between the planet and the outer boundary of the material closer to the star, which tends to move the planet into a smaller orbit. As the years pass, the disk material moves steadily inward, and so does the planet.[10]

Tidal Friction Tends to Halt Planetary Migration

What, then, prevents the planet from spiraling all the way into the star? The answer, which receives less complete approval from astronomers than the basic mechanism of migration (which itself is not totally supported by astronomical experts), has the name of "tidal friction."

How can tides generate friction? Those who visit the seashore have observed what are familiarly called tidal forces, though a more accurate name would be "the effect of differences in gravitational forces." Consider the most famous tidal forces—those that raise the tides in Earth's seas. Both the moon and the sun exert gravitational forces on Earth, and they each attract the part of the Earth closer to them with more force than they do the Earth's core, and they attract that core more strongly than the part of the Earth beyond the core. The differing amounts of force on different parts of the Earth stretch the planet along the direction toward the sun or the moon, making it bulge both toward the attracting object (this seems entirely intuitive) and also away from that object (entirely counterintuitive—but remember that the core feels more attraction than the far side).

Thus differences in the gravitational forces from both the sun and moon make the Earth bulge, the result of what scientists call "luni-solar tides." Because the liquid oceans can react much more easily than the solid land, the seas move against the land to create high and low tides, each occurring twice a day in reaction to a given object's forces on the Earth. Even though the moon exerts far less gravitational force on us than the sun does (so that we orbit the sun, not the moon), the *difference* in the amount of force from place to place tilts the balance in the moon's favor, because the moon has only ¹⁄₄₀₀ of the sun's distance from us. The luni-solar tides therefore basically follow the moon's location with respect to the Earth. Near full moon or new moon, the effects of the sun and moon combine to generate relatively large "spring tides." Alternating with those days, near the first- and last-quarter phases of the moon, we observe less pronounced "neap tides."

Consider, then, an exoplanet that draws ever closer to its star. As it does so, the star exerts tidal forces on the planet, deforming it in a manner analogous to the Earth's deformation by the luni-solar tides. The amount of deformation increases as the planet grows closer to the star, making the planet bulge toward and away from the star. If, as we may reasonably assume, the planet has some rotation, as all the sun's planets do, then the tidal forces on the planet will eventually create "tidal locking," a match between the planet's rotation period and its orbit period. This means that the planet will always present the same face toward the star, just as the moon, for the same reason, matches its rotational and orbital periods, always showing us the "man in the moon." Jupiter's four large satellites have likewise undergone tidal locking, so each of them always shows the same side to any Jovian observers. Because a large object's ability to induce tidal locking of a smaller one varies inversely as the cube of the distance between them, the sun's planets have escaped this fate, although Mercury has long been locked into a three-to-two resonance (see Chapter 9), in which the planet rotates three times during every two orbits around the sun. Once tidal locking arises, the planet will most likely cease its tendency to creep into a smaller orbit. If the planet's orbit were to grow smaller, its rotation rate would have to increase, continuing to match its rotation and orbital periods. But tidal locking keeps the rotation rate fixed, and this in turn inhibits any change in the orbital period.[11]

The reader who has followed this subtle reasoning of how tidal migration can bring a planet just so close, and no closer, to its star may feel rightly congratulated. After this happy moment, the reader, we may hope, will not overly resent the fact that this explanation does not satisfy all planetary-orbit specialists, who can, however, offer an alternative. If both the star and planet possess significant magnetic fields, as, for example, the sun and Jupiter do, then the interaction between these two fields, which grows stronger as the planet approaches the star, may itself inhibit further inward migration. Notice that with this hypothesis, we pass beyond a reliance on gravitational forces to explain all that occurs in the evolution of orbits within a planetary system.

If you don't approve of the attempt to explain how tidal locking can halt the migration of a giant planet, you may happily contemplate the result of failure to stop a giant planet's inward motion and imagine that it falls into its star. You are not alone: Even those astronomers who adopt the tidal-migration explanation find it likely that some giant planets have met this fate, and some have considered how swallowing a planet might affect its star, perhaps to produce a lasting effect that we can still observe. This possibility has led to a research publication with one of the more admirable titles of recent years, "Stars Get Dizzy after Lunch," which proposes that swallowing a planet could cause a star to rotate much more rapidly, implying the possibility of detecting a star's planetary meal *post factum*.[12]

Planets at "Normal" Distances from Their Stars

As we discussed in the previous chapter, exoplanetary systems offer a startling range in their masses and orbits, partially confounding those who expected the solar system to serve as a useful model of the general history of planet formation. If we take the broadest view of the situation, we can state that most planets appear to have formed in approximately their current locations, and that two basic methods of formation remain: accretion, bit by bit though ever more rapidly as growth proceeds, and formation all at once, through instabilities that cause the sudden collapse of a cloud of gas and dust. The most promising representatives of the latter formation process appear in the directly imaged planets described in Chapter 6—comparatively giant objects at large distances from their stars. Smaller objects, and in particular all objects made primarily of rock, can be more easily explained as the result of coalescence of small particles. Both processes become more complex in the presence of magnetic fields, whose strength varies both with time, as a protoplanetary disk contracts to form a star and its planets, and from star to star throughout the galaxy.

If we skip nimbly past such complexities, we can say, with a fair degree of confidence, that astronomers today find the solar system model applicable, in its broadest outlines, to the formation of most other planetary systems, from the planets that nestle close to M stars to the remarkable variety of exoplanets that orbit around sun-like stars. Explanations of why a planet with a particular size, mass, and orbit formed as we find it—or as circumstances have changed it—remain largely a task for future generations. They might begin, for example, with the question posed by the majority of exoplanets we now know: What explains the emptiness of the inner solar system, which contains no objects with orbital periods less than Mercury's 88 days? As shown in Figures 8 and 9 (Chapter 4), a multitude of exoplanets have much smaller orbits and lesser orbital periods. Could some peculiarity of the solar system have prevented planets from materializing close to the sun or removed the planets that did emerge there?

Before we leave this survey of the origin of planetary systems, we should consider one key discriminant among stars: their elemental compositions, which may significantly affect their planets.

Which Stars Are Likeliest to Have Planets?

Soon after the first discoveries of exoplanets, astronomers, who have always been great classifiers and categorizers, began to analyze their data to seek possible relationships between the different types of stars and the likelihood that the stars possess planets. Their classifications hinged on two parameters that distinguish different types of stars. The first of these, which is far more basic, categorizes stars by their *masses,* which determine their sizes and luminosities, while the second, still highly significant, classifies stars by their *composition,* the percentage of each star's mass that consists of elements heavier than hydrogen or helium. The abundances of these other elements reflect, among other items, how far a star has evolved through its nuclear-fusing lifetime. We may consider how each of a star's two key parameters—its mass and its elemental composition—may affect the possibilities of planets orbiting around it.

Astronomers deduce stellar masses from their knowledge of how stars generate energy. Every ordinary star maintains a near-perfect balance between its tendency to contract under its self-gravity and its inclination to explode from the mammoth amounts of kinetic energy released by nuclear fusion. Greater stellar masses bring stronger self-gravitation and more powerful squeezing of the star's interior, which requires larger amounts of energy release to counteract the star's tendency to contract. The star's luminosity (energy released per second) therefore obeys a straightforward rule: Larger masses lead to much larger luminosities (in fact, a star's luminosity varies roughly in proportion to the *cube* of its mass). Our sun, more massive than 95 percent of all stars, outshines them in absolute terms, but its luminosity falls tremendously short of stars that have 5 times its mass and a luminosity more than 100 times greater.

Astronomers can determine stars' masses by measuring their luminosities, which they obtain directly from the stars' distances and apparent brightnesses. Analysis of the stars' spectra opens another avenue to determining stars' luminosities, because the different characteristics of the absorption and emission features in their spectra depend on the stars' luminosities. In addition, the details of the spectra provide important information that distinguishes stars with the same mass: the fraction of their mass that consists of elements other than hydrogen and helium.

Hydrogen and helium, by far the two lightest and most abundant elements in the cosmos, typically provide 99 percent of a star's mass, as they do for the sun. Using a naming system that defies logic and much of history as well, astronomers call *all* these other elements "metals," assigning this label not only to the elements familiarly known as metals but also to carbon, nitrogen, oxygen, and a host of others never tagged with this name beyond the realm of astrophysics. A star's "metallicity" therefore describes the percentage of the star's mass provided by everything *other* than hydrogen or helium.

Objects much smaller than stars, such as the sun's eight planets, have metallicity values far different from the stars'. Like the sun, the four giant planets consist mainly of hydrogen and helium, but they apparently possess sizable solid cores that raise their metallicity fraction to many times the sun's. The inner planets have almost no hydrogen or helium: Though the Earth's extensive hydrogen-oxygen seas may deceive the unwary, they bottom out at depths much less than 1 percent of the distance to the Earth's center. Hence the four inner planets, like their moons, the asteroids, and the trans-Neptunian objects, consist almost entirely of elements heavier than helium. Chief among those elements in the Earth are silicon, oxygen, aluminum, and iron.

Because most of the giant planets' cores apparently formed before they wrapped themselves in thick layers of hydrogen and helium, they, too, like the solid inner planets, owe their formation and existence to the agglomeration of heavier elements that clumped together within the rotating disk of material that gave birth to the sun and its orbiting retinue. When we look outward and seek to find the protoplanetary disks that represent the likeliest sites for planet formation, the straightforward answer seems to be to look for those with the highest metallicities of their central stars. Observational data now supports this conclusion—to an extent.

In 2016, a team of astronomers led by Gijs Mulders of the University of Arizona's Lunar and Planetary Laboratory announced the results of their analysis of 20,000 stars observed by the Kepler spacecraft, for which other astronomers had determined metallicities through analyses of the stars' spectra. Mulders and his colleagues found that closer-in exoplanets occur more frequently around stars with metallicities higher than the sun's, and less often for lower stellar metallicities.[13] The difference becomes most significant for close-in rocky planets; close-in gas planets still display this effect, but less prominently. In contrast, exoplanets at larger distances do not demonstrate a correlation between the frequency with which they appear and the metallicities of their host stars. These results confirm the reasonable supposition that higher metallicities should provide conditions more favorable to forming rocky planets in the face of the evaporative tendencies resulting from proximity to a star.

When we assess how likely we are to discover planetary systems that come close to the model that the solar system presents, we should consider the summary of our current understanding that Stanford University's Bruce Macintosh has made:

I think the most startling aspect [of our discoveries] is just how different all the other planetary systems we've seen [are] from our own. Solar-system twins are hard to detect, so there could still be a significant fraction of them hiding out there, but it's clear that our solar system isn't the only way to assemble one, and in fact it may be a very atypical path: Planets are common, but planetary systems like our own could still be very rare.[14]

One of the other hands in astronomical anatomy and Macintosh's fellow exoplanet hunter, Jason Wright of Pennsylvania State University, has an additional twist on this morsel of truth:

> The search for other places like home has turned into an exoplanetary bonanza—of places that look like anything *but* home. Indeed, the solar system now appears almost unusual in its configuration, and a very *poor* guide to planet formation. But we should be careful not to fall into the same trap a second time: The exoplanets we have found are almost all *more easily detected* than anything in our solar system—today's radial velocity method used on the sun from afar would only [be] sensitive to Jupiter, and the Kepler mission needed to observe over 100,000 stars to find a handful of Earth- and Venus-like planets. Our solar system may yet prove to be the prototype of a common but hard-to-detect outcome of planet formation, and we should be prepared to dust off the old conventional wisdom in case the exoplanets surprise us once again."[15]

Eric Agol offers a broader and more poetic summary of the current situation in exoplanet discovery:

> If every solar system was like our Solar System, we would have found very few planets to date; but we have found them in droves anyway, in the most unexpected places [and] with unanticipated properties. The most startling aspect of the exoplanet field is their ubiquity and diversity: We have detected small planets and enormous planets; young and old planets; planets on small orbits and large orbits; planets orbiting puny stars and massive stars; dense planets, fluffy planets; planets

disintegrating around normal stars, but surviving around dead stars; free-floating planets, and star-hugging planets; some planet systems which are askew, while others are precisely aligned. We couldn't have anticipated this breathtaking menagerie twenty-five years ago, and now that we have the data, it's changing our view of our own solar system, and sharpening our strategies for finding more exoplanets.[16]

For some of us, this leaves only one great task for the unknown future: How and when do we discover life amidst this exoplanetary richness?

12

·

HABITABLE PLANETS AND
THE SEARCH FOR LIFE

A large part of the public and professional interest in the search for exoplanets arises from an innate desire to answer two age-old questions: Do other forms of life exist in the cosmos? If they do, how can we find them and talk with them? Some astronomers caught up in the excitement of exoplanets now regard finding evidence of life on some of these new worlds as the most important goal of planet hunting. "Otherwise, exoplanets are not so exciting," says Abraham Loeb of the Harvard-Smithsonian Center for Astrophysics. "We'd be crazy not to [search for signs of life]," Loeb's colleague Dimitar Sasselov agrees. Olivier Guyon, an astronomer at the University of Arizona, takes it a step further: "I'm only interested in *habitable* planets."[1] Few would challenge the human urges to find potential sites for life and extraterrestrial life itself—but how can we fulfill them?

Two interlocking streams of analysis feed the inquiry into the possible existence of life beyond the Earth. One focuses on an attempt to unravel the mystery of life's origin on Earth, in order to reach conclusions that will allow us to judge which planets or other objects seem likeliest to harbor life. The second concentrates on the search for "biosignatures"—actual evidence, if we judge it so—of life on other worlds. Astrobiologists, as they are now named, therefore seek knowledge of life's origins and its development on Earth so they can decide where best to devote their efforts to find direct evidence of life elsewhere. Let us follow these dual threads toward what we hope may be reliable conclusions that may eventually yield the long-sought demonstration—if this be the truth—that we are not alone.

In any analysis of extraterrestrial life, we must strive to resist two contradictory temptations: (1) to imagine that extraterrestrial life must resemble its terrestrial counterpart in its appearance and biological functions, or (2) to let our speculation roam utterly free, imagining any form of life that may appeal to us. Just how we should use our knowledge of biology and physics to set proper limits on our speculation remains, of course, an open question.

The panoply of known exoplanets underscores the limited nature of our preconceptions concerning the nature of other planetary systems. In a similar manner, our exploration of the solar system has already highlighted the limitations of our ability to assess the likelihood of where life-supporting situations should exist beyond the Earth, and how to find them. These limitations begin with our attempts to resolve two fundamental issues: What is life, and what does it require to exist?

Neither of these questions has received an agreed-upon answer within the scientific community, so any attempt to resolve them requires caution. As to life's definition, we may adopt the majority view that all forms of life must possess the ability to *reproduce* and to *evolve*. Within these familiar verbs—especially the first of them—lurk empires of uncertainty. Consider, for example, that every form of life found on Earth, from the lowliest bacterium to that pinnacle of creation, humanity itself, reproduces itself through the same molecular processes. The omnipresent structure of DNA molecules in living creatures testifies to the unity of life on Earth. But could the first living organisms already have developed these complex double helices? If not, what came before, and how did those creatures reproduce? If so, what path led to anything so complicated?

While preparing to skip lightly over these questions, we may pay homage to ecologically oriented biologists who propose that we should consider life not in terms of individual organisms, or even as kingdoms or phyla of living creatures, but rather as a planet-wide phenomenon, from the perspective that life happens not *on* a planet but *to* a planet. This Gaia-like approach, named after the Greek goddess of the Earth, naturally leads us to ask, in our search for life elsewhere, what conditions on Earth made it possible, even likely, perhaps almost inevitable, that life would originate here? Or did life begin elsewhere, to be brought to Earth in meteoroids or cosmic dust? Happily enough, the question of where Earthlife began presents only a small diversion from the crux of the matter: determining the conditions that favor the emergence of life on Earth from nonliving matter.

Here at least we find some scientific consensus. All living creatures on Earth consist of one or more cells that each encapsulate a bag of water, within which different types of molecules float and interact. Life presumably began with its simplest, single-cell form, but even the multitrillion-celled monsters we call whales and humans are no more than marvelous pyramids of specialized cells, each of which closely resembles its original forebears. Their commonality resides (a bit) in their size and, far more important in the search for life, in their use of water as a liquid conducive to easy interactions among molecules, which float and collide freely and often within every cell.

Supple though our imagination may be, we rightly find it unlikely (at least in the measured opinions of astrobiologists) that solid matter, with its firm interlocking of atoms and molecules, can provide a useful matrix within which a multitude of molecular interactions might lead to life. Gases, on the other hand, could serve well as the sort of medium we require, as they offer none of the resistance of solids. However, because gases are almost always much more rarefied than liquids, the concentration of molecules floating in a sea of gas will almost always fall far below what occurs within a liquid. Through this straightforward, some might say simplistic, analysis of the fundamental three stages of matter, we—as others did before us—tend to conclude that life most likely originates, as Earthlife clearly did, within a liquid bath or solvent.

Liquids do have two drawbacks for providing the universal solvent for life. First, they apparently require a solid surface on which to rest; otherwise they become separate, tiny droplets that slowly evaporate. Second, any substance that forms a liquid does so only within a restricted range of temperatures, beyond which it either freezes into solid or evaporates into gas. This range varies among various substances, and each substance also depends on the ambient pressure. Thus water, for example, remains liquid between 0 and 100 C when subjected to the atmospheric pressure at the Earth's surface, but it evaporates at lower temperatures in lower-pressure situations.

We have thus arrived, perhaps all too readily, at establishing life's universal requirement of a liquid, and thus for a solid surface that maintains appropriate, liquid-inducing pressures and temperatures. Before human exploration of the solar system, only two objects appeared to satisfy these requirements: Earth and (barely) Mars. In fact, liquid water cannot exist for long on Mars under today's conditions, because even though Mars's surface temperature occasionally rises above 0 C, the atmospheric surface pressure nowhere exceeds the minimum required to keep liquid water from quickly evaporating. Although evidence shows that water exists temporarily on Mars, most experts on extraterrestrial life (often described as a subject without a subject matter) have found themselves imagining that pools of water permanently exist beneath the Martian surface—still a promising possibility, but one not yet investigated by even our most advanced rovers.

The Habitable Zone and Its Limitations

Meanwhile, the widely accepted conclusion that life most likely requires a liquid bath for its origin and development leads to some straightforward direction in searching for life beyond the solar system. We must look either to extraterrestrial objects that have internal sources of heat that allow subsurface oceans to exist, or to planets that orbit within the "habitable zone" of their stars, a term coined by the astrophysicist Su-Shu Huang in 1959.[2] The habitable zone denotes the Goldilocks distance from its star that allows the planet to maintain a temperature favorable to a liquid, and thus to life, on its surface. Astronomers continue to argue about the precise definition of the habitable zone, because we cannot yet specify accurately the conditions required for the origin, existence, and maintenance of life on other worlds. Nevertheless, the phrase "habitable zone" now refers to the range of distances from a star within which an orbiting planet can maintain surface temperatures at which water remains liquid, leaving the subsidiary question of an atmospheric surface pressure for further discussion.

Calling this region the habitable zone often embraces the belief that liquid water, rather than another compound, provides the medium within which complex molecules can float and interact. We may, of course, imagine other common substances, such as ammonia or methyl alcohol, that might provide a substitute for water as life's primary liquid. Planets' habitable zones will acquire different boundaries if we assign different liquids the key role of providing the basic solvent for life. However, we may conclude, in line with most reasoning on the subject, that water seems in many ways to provide the best solvent, most notably because of its abundance, but also for some more subtle qualities, such as its surface tension and the fact that the solid form of water floats on top of the liquid form. In general, we lose only details if we choose to focus on water as life's likely fundamental liquid.

In the solar system, the habitable zone based on liquid water spans the region from the orbit of Venus out to the orbit of Mars. Devoid of a thick atmosphere, Venus might be able to maintain liquid water on its surface, but the actual Venus, wrapped in a stifling blanket of carbon dioxide 100 times thicker than Earth's nitrogen-oxygen gauze, bakes at temperatures close to 600 C. Although the temperatures on Mars almost stay below 0 C in almost all regions and seasons, the possibility of liquid water does arise temporarily, but the thin atmosphere provides too little pressure to prevent nearly immediate evaporation.

If extraterrestrial life depends on water, a star's habitable zone will extend from an inner border, where water boils at 373.16 K (100 C), to an outer border, where the temperature never rises above the freezing point of water, 273.16 K (0 C). The star's luminosity governs the dimensions of the habitable zone, but we should also recall that its inner boundary will depend on the boiling point of water, which in turn depends on the atmospheric pressure on a planet's surface. For a sunlike star, the habitable zone extends roughly from the orbit of Venus to a distance comparable to the orbit of Mars, or from 0.7 to approximately 1.6 AU. (The fact that the boundaries correspond to the orbital distances of actual planets may be regarded as sheer coincidence.)

If we conclude that a substance other than water may furnish life's basic bath, the different temperature ranges at which the substance remains liquid will change the boundaries of the habitable zone. Although ammonia and methyl alcohol have freezing and boiling points different from water's (195 K and 243 K for ammonia, 176K and 338 K for methyl alcohol), their role as life's basic liquid would shrink or enlarge the dimensions of a star's habitable zone only by fairly modest amounts.

In contrast to the fairly generous extent of the habitable zone that surrounds a sunlike star, consider the habitable zones that surround the most common stars in the Milky Way, the M stars, which typically have only $1/100$, or even less, of the sun's luminosity. The laws of physics require that if we seek the same temperatures as those in the vicinity of a sunlike star, we must approach an M star that has 1 percent of the sun's luminosity to a distance just one-tenth of the distance appropriate to the sun's neighborhood. Before the era of exoplanets, few astronomers seriously considered that these comparatively tiny habitable zones might include one or more planets. Today we know how wrong we were: Many planets orbit much closer to their stars than the sun's planets do.

Looking Beyond the Habitable Zone

Fifty years ago, scientific speculation regarded the habitable zone as a fundamental determinant of potential extraterrestrial life. The exploration of our planetary system seriously altered this assessment. On the one hand, recent images from the Mars Reconnaissance Orbiter have provided evidence that liquid water may exist on the surface of Mars, though only temporarily and in a few situations. This evidence consists most notably of streaks apparently formed by water that flow down a host of downslopes on Mars during the warm season. These recurring, temporary flows could consist of brines, water rich in various salts, that freeze at temperatures dozens of degrees below 0 C. Even more recently, however, a new interpretation of these streaks ascribes them to flows of sand and similar granular materials that are unaccompanied by any liquid water.[3]

If nature indeed sets the traditional limits on a star's habitable zone by requiring temperatures above 0 C for liquid water, then it turns out not to set these limits by requiring that an object's *surface* temperature rises above this mark. Our amazing automated explorers of the solar system have found two water-bearing worlds, Europa and Enceladus, which orbit the sun far beyond the realm of the terrestrial planets. In orbit around the giant planets Jupiter and Saturn, respectively, these worlds remain so far from the sun that no liquid water could exist on them. Nevertheless, they possess worldwide oceans *beneath* their surfaces.

How can these modest objects, each about the size of Earth's moon, maintain these comparatively enormous seas in liquid form? The answer arises from their internal heat, the result of constantly changing stresses within the satellites' interiors. The stresses arise from changes in the amount of gravitational force felt by Europa, both from Jupiter's gravitational attraction as Europa moves along its slightly elliptical orbit, and from the planet's nearby large moons as their distances from Europa change. The greatest tidal stresses induced by these effects arise on Io, Jupiter's innermost large moon, where internal heating causes giant volcanic plumes of sulfur-laden gas to erupt from Io's surface. Io, however, has no ocean beneath its constantly recoated exterior.

On Enceladus, similarly variable gravitational forces from Saturn and its neighboring large moons likewise induce internal stresses that heat the moon's interior and maintain a liquid ocean beneath the surface. In 2015, the Cassini spacecraft, which surveyed Saturn, its rings, and its myriad satellites throughout the dozen years before its mission ended in September 2017, observed plumes of hydrogen gas that periodically erupt from Enceladus. On Earth, hydrogen molecules appear when water reacts with iron-bearing minerals in hydrothermal pools. This hydrogen allows microorganisms in these pools to derive their energy from a chemical reaction called "methanogenesis," a reaction of hydrogen with carbon dioxide that generates methane gas. A metabolism based on this sort of chemical reaction requires no dependence on the energy from a star. Terrestrial biologists have found methanogenic bacteria that rank among the oldest life forms on our planet, and continue to thrive in the deep ocean, or beneath kilometers of ice, in environments to which sunlight never penetrates. The hydrogen plumes on Enceladus lend support to the possibility that methanogenic forms of life could thrive even now on Saturn's sixth-largest satellite.

Titan, Saturn's largest moon, which is 5 times farther from the planet than Enceladus, has 10 times Enceladus's diameter and more than 1,000 times its mass, but no significant source of internal heat. Titan nonetheless maintains liquid on its surface, in pools and lakes at various locations around Titan's extensive surface, whose total extent nearly equals the land area on Earth. Titan's lakes, however, consist not of water but of methane, whose hydrocarbon molecules (CH_4) can remain unfrozen even at Titan's surface temperature of 94 K, close to 200 K below the freezing point of water. Speculation about life in methane lakes has a long history, and NASA has dreams of sending probes that would sail the Titanian seas to search for life—not methanogenic but methane-based, as Earthlife has always been entirely water-based.

Life on Exomoons?

Our recent discoveries about conditions on Europa, Enceladus, and Titan emphasize an important conclusion in the quest for extraterrestrial life: We should not restrict our conjectures about life to *planets* similar to Earth and Mars, but instead we should include the moons that may orbit those planets, and in particular the satellites of giant planets. As we discussed at length earlier in this book, giant planets appear in the exoplanet catalogs in impressive numbers, and some of them are so close to their stars that life might prove impossible on any of their moons. But there are many more giant planets at much greater distances that definitely allow for the possibility that their satellites, which we may call "exomoons," could resemble the three solar-system moons that we have discussed. Prospects for exomoon detection center on three methods—direct observations of transits, transit timing, and gravitational microlensing—all of which require improvement before they yield their target.

Transit observations have already found planets with about half the Earth's diameter, and before long they should prove capable of finding objects with one-third or even one-quarter of our planet's diameter, thus reaching the size realm of our moon and of the solar system's six larger satellites. Sufficiently accurate observations of exoplanet transits could reveal the tiny diminutions in starlight that a planet's moon would cause. The appeal of this approach has led to a continuing reanalysis of the Kepler data on planetary transits. For now, we can state that the majority of transiting exoplanets do not have satellite systems comparable to Jupiter's four large moons.

Exomoons would induce tiny but measurable changes in the times of planetary transits, thanks to a moon's tug on its planet in different directions through the course of an orbit. Similar timing variations served astronomers marvelously well in determining the masses of the seven planets around TRAPPIST-1 (see Chapter 11). To extend this method to objects much less massive than the TRAPPIST-1 planets will require observations with one or more of the future observatories described in the following chapter.

Gravitational lensing could reveal planetary moons, if we could detect not only the secondary blip that arises from a planet that orbits its star, but also a tertiary blip that would signal the existence of an object significantly smaller than the planet. If the secondary and tertiary blips almost coincided in time, astronomers might be tempted to assign the tertiary one to the planet's moon, but the possibility would remain strong that instead the tertiary blip arises from a chance line-up with another planet much less massive than the one that produced the secondary blip. Once the WFIRST spacecraft described in Chapter 13 begins to amass a collection of detailed microlensing observations, this conclusion can be sharpened, perhaps to find that a minority of these planets do possess one or more large satellites.

If all else fails—if we conclude, against the accumulating evidence—that liquid water may appear only rarely on exoplanets and their moons, we may fall back on the speculation that molecules other than those usually proposed as life's solvent might play this role after all. Before we find ourselves reduced to this sorry pass, let's take a look at the great areas of speculation for water-based life on the newly discovered host of exoplanets around underluminous red dwarf stars.

The Hazards and Advantages of Living around a Star

Life on our planet depends almost entirely on the sun's steady energy output, which has shown an impressive constancy over the past millions of years. The exceptions to this statement appear primarily in short-term outbursts known as "solar flares," sudden eruptions from regions near the sun's surface. Solar flares often create "coronal mass ejections," or CMEs, typically accompanied by intense x-ray emission, that shoot streams of charged particles into space. CMEs pose a definite hazard for our life on Earth, not from a significant increase in the sun's brightness, which does not occur, but from the ejected particles that reach the Earth a few days later. These protons and electrons can disrupt radio communications when they strike the ionosphere, the atmospheric layer rich in charged particles that acts as a reflector for radio waves.

A giant CME could do immense damage, not only to our civilization's ability to communicate but to all electrical devices as well. We have historical records of events such as these, though they have occurred (so far) before modern communications and electronics came into vogue. In the middle of September 1770, bright auroral displays appeared around the globe for nine consecutive nights, as streams of charged particles from a solar CME, guided by the Earth's magnetic field, encountered atoms high in the atmosphere.[4] Unlike almost all such displays, which last for only one or two nights, the 1770 auroras testify to an immense and long-lived solar outburst, perhaps one even more powerful than the mighty solar CME that produced the better-known "Carrington event" in September 1859. Named after the astronomer who actually saw the solar eruption that sent charged particles toward the Earth, the Carrington event generated auroral displays as far south as Cuba, some of which were so intense that many people believed their cities to be on fire. The flood of fast-moving charged particles also disrupted telegraph communications (the "Victorian internet") around the world.[5]

Today a Carrington-like event would wreak havoc to the tune of $2 trillion (according to an estimate by the National Academy of Sciences). Aside from rendering inoperable many types of electric equipment, including the pumps that provide water to homes and businesses, such a blast could fry giant transformers that would take years to replace.[6] Just such a mighty event occurred on July 23, 2012, when the geometry of the solar system gave Earth a lucky escape. Each of the sun's giant CME ejections sends floods of charged particles in a particular direction, and on that day, the sun's rotation had directed its blast not at the Earth but about

90 degrees away, toward a location with a sun-orbiting satellite that recorded its magnitude.

If a Carrington-like eruption happens perhaps once each century, then longer time intervals probably bring still more powerful blasts. The most energetic of these could remove a planet's ozone layer, raise the surface temperature by tens of degrees or more, and strike the surface with ultraviolet radiation capable of destroying most forms of life. The Harvard astronomers Manasvi Lingam and Abraham Loeb have considered the possibility that at least some of the mass extinction events on Earth, which the fossil record shows to occur every few tens of millions of years, might have arisen from the sun's most powerful eruptions.[7]

What protection can we imagine against the more modest, but still mighty CMEs that may erupt about once each century, and seem likely—perhaps almost certain—to appear during the current millennium? Lingam and Loeb have suggested that we might construct a magnetic deflector to keep the floods of charged particles away from Earth. If their rough estimate of $100 billion to create such protection seems expensive, we may note that these astronomers estimate potential future damage from a giant CME to reach at least 100 times that amount.[8]

When we turn our attention to the many M stars that have planets close to them, we find that the problem of stellar flares grows larger. In comparison to sunlike stars, M stars tend to generate more extreme flares, and do so at shorter intervals. A week-long survey of 215,000 M stars made with the Hubble Space Telescope in 2010 found that 100 of them had sudden increases in their energy output by 10 percent or more that lasted for about 15 minutes.[9] Extrapolation of these numbers suggests that a representative M star undergoes such a flare every few decades. In April 2014, a satellite studying the cosmos in x-rays detected a flare from an M star 60 light years away, whose luminosity ordinarily equals $\frac{1}{500}$ of the sun's. The flare temporarily created a flow of energy 10 times the sun's in visible light, and 10 times more than this in x-rays.

Outbursts like these pose serious threats to the viability of life on a planet around an M star, the more so as still more violent flares would likely appear as longer intervals of time go by.[10] But before we incline toward the conclusion that life on the exoplanets of M stars might prove nearly or totally impossible, we should note that some M stars, especially the older ones, may provide more stable conditions, more favorable to life. In 2017, astronomers used the HARPS spectrograph (see Chapter 4) to make radial-velocity observations that revealed a planet around the M star known as Ross 128, just 11 light years away, a modest distance that makes Ross 128 the twelfth closest star system to the sun.[11] (For those interested in our immediate astronomical future, Ross 128's motion toward the solar system will make it our closest starry neighbor about 70,000 years from now.) Its planet, Ross 128 b, whose minimum mass equals about 1.35 Earth masses, now ranks as the closest known exoplanet after Proxima Centauri b. Orbiting its red dwarf every 10 days, at a distance from Ross 128 only 5 percent of the Earth–sun distance, Ross 128 b receives about one-third more stellar energy per second than the Earth does, qualifying it as potentially habitable, even though it's probable that tidal locking would always leave one side in starshine, and the other in darkness. In its mass, orbital size, orbital period, and stellar energy received per second, Ross 128 b provides a near twin to Proxima b (see Chapter 4).

What cheers life-hunting researchers even more than Ross 128 b's proximity to Earth, however, lies in the fact that Ross 128, with an age estimated between 5 and 10 billion years, belongs to the category of old red dwarfs. In rough harmony with humanity's characteristics, aged red dwarfs lead a much quieter life than their frisky young cousins, as they emit fewer and less powerful flares as time goes on. Proxima Centauri has an age similar to the sun's—a bit less than 5 billion years—but Ross 128's (probable) significant great age implies an even more quiescent star. This dim red dwarf and its newfound planet remind us that even though life may have had a tough (or impossible) time on M stars' planets during the universe's earlier years, the passage of 5, 10, or even more billion years may have allowed, or may eventually allow, conditions on many of these planets suitable for life to originate and to flourish, possibly (as apparently happened with early life on Earth) beneath the planetary seas that we can imagine along with life itself.

As long as we are absorbing the long, long view of life, we should salute the future of life that may exist close to red dwarf stars for the fact that these stars have one key advantage over sun-like stars: their immense lifetimes. Stars similar to the sun have total lifetimes measured in billions of years—around 10 billion for the sun itself. M stars, the tortoises of the starry realm, last for hundreds of billions or even trillions of years. Speculation about civilizations older and wiser than our own usually confine themselves to head starts measured in thousands or millions of years, but the unshackled mind can easily extend these conjectures to societies that could achieve lifetimes tens or hundreds of times greater than the current age of the solar system.

Biosignatures

How can we find evidence for life on planets many light years from Earth? The most promising approach lies in analyzing a planet's atmosphere, searching for chemical compounds that arise from life and cannot otherwise be produced—according, of course, to our current and best understanding. This search will begin in earnest with the James Webb Space Telescope (see Chapter 13), which should provide astronomers with the first instrument capable of making a detailed spectral analysis of the atmospheres of Earth-like exoplanets. Soon after the new space telescope, usually called the JWST, begins to amass data, three great Earthbound telescopes of the future, each of them with mirrors 30 meters in diameter (also described in Chapter 13) should provide complementary spectroscopic observations.

Two aspects of exoplanet observations favor success in measuring the atmospheric composition of exoplanets tens or hundreds of light years from Earth. We should observe planets in transit, so that we can subtract the star's spectrum from the total spectrum observed of the star plus the planet (see Chapter 5). In addition, we must, at least initially, observe low-luminosity M stars, whose small sizes make exoplanet transits easier to observe in detail. More precisely, we would like to observe an Earthlike planet as large as we can find, in orbit within the habitable zone of a star with the lowest reasonable luminosity and size. Even so, teasing out the details of the planet's atmospheric composition will offer a formidable challenge, requiring hundreds or even thousands of hours of JWST observations. The most promising areas for this effort reside in the infrared portion of the spectrum, for a number of reasons. First, planetary atmospheres show their greatest richness of spectral features, typically arising from molecules of water, methane, and carbon dioxide, in the infrared. Second, infrared radiation typically penetrates dust clouds that block shorter-wavelength visible light, obscuring the spectral features that we seek to detect. Third, infrared observations have the advantage of allowing astronomers to sample the entire atmosphere of an exoplanet by observing what happens when the planet disappears behind the star (see Figure 13).

What spectral features might provide the biosignatures that astronomers seek? The classic example suggested by life on Earth consists of finding coexisting oxygen and methane molecules. Methane (CH_4) comes from a variety of creatures, both while they

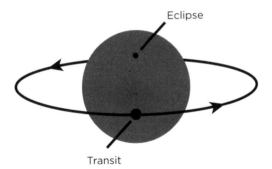

Figure 13 If we observe an exoplanet moving in an orbit that coincides with our line of sight, we observe both a transit, when the planet passes between us and the star, and an exoplanet eclipse, when the planet passes directly behind the star at the opposite point of its orbit.

are alive and while they decay, but also by nonbiological pathways (the giant planets' atmospheres consist largely of methane). Oxygen molecules (O_2) likewise arise from varied biological and nonbiological pathways. However, oxygen reacts strongly with methane to yield carbon dioxide and water. The fact that the Earth's atmosphere simultaneously contains both a large amount of oxygen and a smaller but significant amount of methane demonstrates that some mechanism continuously releases methane into the air. Measurements show that terrestrial methane now arises primarily from livestock flatulence and other bovine products, a tribute to ruminants' abundance, mainly in feedlots and pastures, in the modern world.[12]

Since we can hardly expect extraterrestrial forms of life to follow the same metabolic pathways as Earthlife does, we would be foolish to conclude that the search for life elsewhere should concentrate on finding methane and oxygen in the same place. (It wouldn't hurt to keep looking for them, of course.) Further analysis will help to refine our goals in seeking biosignatures, and finding even one form of extraterrestrial life would provide a giant step toward establishing the broad procedures to follow as we search. We encounter little chance of error in stating that the physics of astrobiology will be the same as that on Earth, but the biology will almost certainly be different.

When can we expect to succeed? Soon, we may hope. A poll taken at a meeting of astrophysicists and astrobiolgists in 2017 asked the question, "Will scientists find evidence of life on an exoplanet by 2040?" Fewer than 40 percent of the respondents gave an affirmative answer, though naturally this fraction increased upon extension of the timeframe by 10 to 20 years.[13]

All who attended this meeting would agree that polling offers no substitute for investigation. From the properties of exoplanets already studied, "we've learned to expect the unexpected," says Shawn Domagal-Goldman of NASA's Goddard Space Flight Center. Goldman, about to turn 40, has pinned his career to the day when, as he puts it, "we start to explore the astrobiological nature of these worlds in the same way as we've explored their astronomical natures." And how long will that take? As a lifelong Cubs fan, Goldman inclines toward optimism, figuring that in 15–20 years, or maybe a bit longer, we should be able to determine which exoplanets support life.[14] At that point, we can recycle previously written books, if they still exist on paper, and write a new chapter in our hunt for life in the universe.

13

·

FUTURE APPROACHES TO HUNTING EXOPLANETS

n astronomical research, as in other areas of life, success fosters attention; attention stimulates funding; and funding brings—more often than not—future triumphs. In considering the many ways to find exoplanets, we can follow several avenues of differentiation and categorization. We might focus on location, discriminating among Earthborne, spaceborne, or (as we dream further into the future) interstellar-borne observatories. Another approach, which we will follow in this chapter, centers on the planet-finding techniques that we have encountered: the radial-velocity, transit, and other methods, including improved attempts to image planets directly. As has been true in the past, some of these avenues toward success interact with, as well as complement, one another. In the interest of simplification, however, let's proceed along our now well-known pathways in an attempt to peer one or two decades into the future.

Soon to Come: The James Webb Space Telescope

Before the end of 2020, if all goes well, the James Webb Space Telescope (JWST), the successor to the amazingly successful Hubble Space Telescope, will finally achieve its orbit, and a year later it will have passed all tests and acquired full functionality. Once known as the Next Generation Space Telescope, the JWST received its new name in 1992 in honor of James Webb, who led NASA throughout most of the 1960s, as the manned space program Apollo passed from its earliest stages into final preparation for the launches that sent 12 men to the moon from 1969 through 1972. Twenty-two different countries have contributed to the creation of the JWST, which represents a great leap forward from the Hubble Telescope and will provide striking new views of the cosmos. Every aspect of astrophysical research should profit from the wealth of observational data obtained by the JWST—if it works!

This caveat, always an appropriate one for complex systems sent into orbit or on long trajectories through the solar system, applies in spades to the JWST, whose history includes long delays in design and construction, along with enormous cost overruns. The most fundamental problem confronted by the telescope's developers lay in their desire, indeed their need, to create a space telescope with a mirror much larger than the Hubble Telescope's 2.4-meter span, the largest that could be carried into orbit on the Space Shuttle.

JWST's 6.5-meter mirror has more than seven times the collecting area of the Hubble Telescope's, allowing it to capture radiation from fainter sources than the Hubble Telescope ever could. But how could such a large mirror be launched without the need for a launch vehicle much larger than the Space Shuttle? The JWST's designers resolved this apparent dilemma by creating a mirror composed of 18 segments, formed from light-weight beryllium metal coated with gold; the segments are stacked atop each other for launch and ready to snap into place, one edge grasping its neighbor, once the segments become weightless in orbit. The entire mirror weighs less than the conventional low-expansion glass of the Hubble Telescope's much smaller reflector. The greatest challenge in executing this novel design lay in the fact that the "snap-into- place" mirror segments must achieve an alignment better than a fraction of a wavelength of the radiation under observation—that is, better than about $1/10,000$ of a millimeter. All of the segments of the JWST's main mirror, as well as the secondary mirror that receives the beam from the primary and reflects it toward the spacecraft's instruments, have actuators, small motors that help the mirrors achieve near-perfect curvature, behind them.

The next and almost equally basic problem facing the JWST sprang from its primary goal: observing the cosmos in infrared radiation. The rationale for infrared observations has a twofold motivation. First, our attempts to observe the early universe consist of our examination of radiation that has been traveling through space for 10 billion years or even more, carrying with it the imprint of how things were within a few billion years of the big bang. The continuing expansion of the universe has shifted the wavelengths with which this radiation now reaches us, so that radiation once emitted as visible light or ultraviolet radiation now arrives as infrared. Cosmologically oriented astronomers therefore conceived of the JWST as the mighty instrument that would unveil the early universe, which remains inaccessible in the visible-light and ultraviolet forms of radiation that were the Hubble Telescope's specialty. Other astronomers have eagerly anticipated the JWST's ability to examine infrared-emitting sources throughout the Milky Way, which include the coolest stars and, arguably even more important, even cooler objects such as exoplanets. Instead of visible light, such objects—including planets, stars, asteroids, comets, and indeed all objects smaller than stars—radiate primarily infrared, the typical output of any object with a temperature in the range from a few tens of degrees above absolute zero to the low thousands of degrees. The Earth, for instance, with an average surface temperature close to 290 K (about 35 K higher than it would be if our planet lacked an atmosphere), emits profuse amounts of infrared, but—except for human creations—almost no visible light or ultraviolet.

Therein lies the second serious problem, which the Spitzer telescope, like the JWST, has had to overcome: Any spaceborne infrared telescope must compete with a host of cosmic sources, of which the closest and brightest interfere with attempts to study the farther and fainter ones. No nearby object radiates more strongly than our planet, a fact that barred the Hubble Telescope, even before it was launched, from making observations at all but the shortest infrared wavelengths. Unable to pass beyond the 300-kilometer altitude at which the Space Shuttle left it, the Hubble Telescope had no real chance to study other objects' infrared emission, and it was not designed to try. (Although its low orbit removed this opportunity, it also offered the Hubble Telescope the possibility that astronauts could return with the Space Shuttle to repair and upgrade the instrument; this in fact rescued the telescope from the uselessness to which its misshapen mirror seemed to doom it, and it also allowed astronaut missions to prolong and to increase the telescope's usefulness throughout the past two decades.)

The JWST's designers decided to send their new and improved telescope to the Earth's L2 point, a location along the extension of the line joining the sun and Earth, 1.5 million kilometers beyond the Earth, that restricts the natural tendency of any object that orbits the Earth to drift with time into a new position.[1] But even at this favored location, which is 50,000 times farther from the Earth than the Hubble Telescope and nicely placed for communication with the Earth because it always lies in the direction opposite to the sun, the spacecraft's infrared-observing capabilities faced serious impairment from the heat released by the Earth, and far more by the sun.

The JWST will deal with this problem by unfurling a five-layer sunshade 12×20 meters in size (Figure 14). The layers in this amazing sunshade, made from a supertough material called Kapton, are each only $\frac{1}{40}$ of a millimeter thick. (To be precise, four of the layers have this thickness, while the sun-facing fifth layer has double this thickness.) Each of the membranes captures some of the heat from the telescope and radiates it out the sides of the sunshade, while the four vacuum spaces between them provide excellent insulation. The sun-facing sheet carries a special coating to reflect solar radiation, and the net effects from its five-sheet sunshade will allow the JWST to radiate its heat into space. The temperature on the sun-facing side of the sunshade will remain near a broiling-hot 360 K, whereas the dark side will cool to just about 40 K. At this temperature, the telescope will radiate sufficiently small amounts of infrared to allow one of its two detectors, which senses shorter-wavelength infrared radiation, to function successfully without additional intervention. The other, longer-wavelength infrared detector, which must receive further attention to avoid overwhelming interference from the JWST itself, will be cooled by an innovative closed-cycle refrigerating system to a temperature less than 7 K.

At some point in 2019, the JWST will—if it remains on its current schedule—leave the Centre Spatial Guyanais near Kourou in French Guiana, the spaceport for the European Space Agency (ESA), from which ESA's heavy-lift Ariane 5 rockets, the agency's workhorse, have made more than 80 consecutive successful launches. The rocket and launch represent a key contribution to JWST made by ESA, the European counterpart to NASA, whose

Figure 14 This photograph of a full-scale model of the JWST, which demonstrates the size of its mirror and sunshade, includes some of the scientists, engineers, and technicians responsible for its design and construction at NASA's Goddard Space Flight Center in Greenbelt, Maryland. During the launch, the segmented mirror will fold the three-segment portions on its left and right sides at 90-degree angles so that the mirror will fit into the payload bay of an Ariane-5 rocket, and the sunshade will be furled into a pocket below the mirror. (Courtesy of NASA)

22 member nations, spanning Europe from Ireland to Romania, have created this collaboration to explore of the cosmos. In addition, ESA created one of the telescope's spectrometers, which will measure shorter wavelengths of infrared radiation, and it collaborated in building a mid-range infrared camera and spectrometer. The Canadian space agency has built the JWST's star tracker and an instrument to obtain images and make spectrometric measurements in the shortest infrared wavelengths.[2]

Five days after launch, having reached a distance more than half a million kilometers from Earth, JWST will begin the first of its two most difficult and tension-filled maneuvers. At that point, the telescope's automated system must spread its five layers of

$1/_{1,000}$-inch-thick fabric without tearing any of its layers or having them stick together, in order to create a sunshade with an area of 240 square meters designed to last for as many years as the JWST observes the cosmos. Six days later, the hexagonal mirror segments, standing high in groups of three or four as JWST left Earth, must swivel downward and lock together in order to form the mirror's single 6.5-meter surface.

Compared to the two great challenges posed by the sunshade and the mirror, the JWST's other problems may have appeared merely worrisome but nevertheless proved serious. During the early 2000s, as the JWST's projected launch date seemed to slip by one year with every passing year, the telescope's projected cost rose from an initial $1.6 billion to the vicinity of $9 billion. Throughout the mid- and later 2000s, engineers solved problem after problem, so that in 2011 NASA could finally set a firm launch date—which has slipped so far by only two years. If the JWST achieves successful orbital entry in 2020, more than 30 years of enormous effort will have reached a culmination.

JWST's two infrared cameras will detect and study an enormous number of faint infrared sources, many of them near the limits of the observable universe. From an exoplanetary perspective, however, the new space telescope's most important feature will reside in its infrared spectrometer, capable of analyzing the different wavelengths of infrared radiation in fine detail. Because these details capture the signatures of the molecules that emit or absorb radiation of a particular wavelength or set of wavelengths, spectroscopic analysis offers the prospect of determining in detail the composition of the gas that causes emission or absorption.

Astronomers have already adopted this method in their studies of transiting exoplanets (see Chapter 5), but the JWST will bring this approach to a new level. Even though the long lead time toward the JWST's operation in space has meant that the telescope's designers did not concentrate as fully as they might have on the potential for observing exoplanets (whose numbers and properties have steadily gained astronomical significance), the JWST offers an excellent instrument not so much to find transiting exoplanets on its own as to study in great detail transiting exoplanets, and their atmospheres in particular, that other spacecraft have discovered. As we discussed in Chapter 4, by comparing detailed spectroscopic observations of a star during planetary transits with observations made at other times, astronomers can deduce the details of the planet's atmosphere, which adds to, or subtracts from, the features of the star's own spectrum. The JWST's ability to observe over a wide range of infrared radiation should reveal the presence and the abundances of the molecules most likely to exist in exoplanet atmospheres.

Seeking Exoplanet Images with a Coronagraphic Mask

In its detailed observations of exoplanets, JWST will employ a coronagraphic mask (discussed in detail later in this chapter) that blocks starlight in order to allow detection of much fainter planets close to the star. This arrangement should enable the telescope to obtain a direct view in infrared radiation of young Jupiter-like planets in orbit. The Hubble Space Telescope has a coronagraphic mask similar in concept, but it could never deploy the mask to obtain exoplanet images, partly for technical reasons but also because of a key difference between observations in visible light and those in the infrared domain—which the Hubble Telescope could barely explore. Recall that a distant observer of the solar system, for example, would find that the sun outshines its largest planet, Jupiter, by about a billion times in visible light, but only by about a million times in infrared. The reduction in this ratio by a factor of 1,000 allows the possibility of coronagraphic observations in the infrared. For younger Jupiter-like planets, which generate more infrared radiation than Jupiter does, the contrast ratio might decrease from a million to a mere 100,000.

The reader should be cautioned once again not to be fooled by the use of the word "image," which to most people connotes a picture with some detail. Even the JWST cannot hope to do more than to record large exoplanets as a single bright dot. The hope of obtaining full-blown images that might show, for example, the contrast between oceans and continents on an Earthlike planet rest with far more advanced systems, long discussed and to some extent designed, that await further progress and funding.

Those who feel that we must pick up the pace in our efforts to obtain images of exoplanets can examine the proposal available at projectblue.org and decide whether they wish to support this initiative to create and launch a lightweight space telescope designed to photograph any and all planets in the Alpha Centauri system (see Chapter 3). Projects such as this one testify to the public's ongoing interest in astronomy: One rarely finds groups of amateur chemists (at least legal ones) or amateur physicists (at least scientifically grounded ones), but amateur astronomers around the world provide important moral and, at least indirectly, financial support in our continuing efforts to understand the universe.

The Gaia Spacecraft and the Rise of Astrometry

One of the great astronomical attempts to map our cosmic environment now proceeds with remarkable success, but the steady and boring nature of this effort leaves it unsung by science popularizers and nearly unknown to the public. This enterprise resides in a spacecraft 1.5 million kilometers from Earth, circling the sun near the same L2 point of the Earth's orbit around the sun that the JWST will occupy (see note 1 to this chapter). In

December 2013, the European Space Agency launched the Gaia spacecraft toward this position, and since 2014 Gaia has performed to near perfection throughout the initial stages of its nine-year, billion-dollar mission.

Gaia's most basic operation, which has been fundamental to astronomy since humans first recorded what they see in the sky, consists of making repeated measurements of stars' positions on the sky. (Gaia also observes individual objects within the solar system, as well as galaxies far outside the Milky Way, but for now we will gloss over these operations.) In the calm of interplanetary space, with a large sunshade to maintain its thermal equilibrium, Gaia uses a billion-pixel camera (about 100 times more pixelated than the camera in a smartphone) to record the positions of nearly a billion stars, over and over again, with a precision far better than any obtainable beneath the Earth's atmosphere. For the brighter stars that Gaia observes (still only $1/10,000$ as bright as the faintest stars visible to our unaided eyes), this accuracy reaches 20 millionths of a second of arc—5.5 *billionths* of a degree! For stars fainter than this—up to 100 times fainter—Gaia's positional accuracy diminishes to one-tenth of this value. Gaia's repeated measurements allow astronomers to determine the distances to these stars by measuring the yearly apparent displacement in position, which varies in inverse proportion to these distances, that arises from Gaia's yearly motion around the sun. This allows Gaia to create a three-dimensional map of the distribution in space of the sun's neighboring stars. In addition, Gaia's measurements record the stars' "proper motions," meaning their individual trajectories through space as they all orbit the center of the Milky Way galaxy in which we live (see Chapter 3).[3]

And there's more: Gaia fulfills astronomers' long-held dreams of finding exoplanets through astrometry, the repeated measurement of stellar motions to search for the sinusoidal deviations from straight lines that arise from one or more orbiting planets. As we outlined in Chapter 3, a planet's gravitational force on its star will make it deviate measurably from what would otherwise be its steady, straight-line trajectory through space. Gaia's observations of hundreds of thousands of stars can reveal planets with masses comparable to Jupiter's, orbiting within about 2 AU of their stars, with this astrometric technique.

Why, then, have we learned nothing of Gaia's list of planets orbiting other stars? The astronomers who designed and now operate Gaia followed the long-term planning that characterizes the collaborative enterprise of the European Space Agency. ESA's first data release in 2016 provided the positions and brightnesses for just over 1 billion stars, and the distances and proper motions for about 2 million of them, along with measurements of the changing brightnesses of variable stars. Later data releases will include the repeated measurements needed to identify those stars with orbiting planets, because the planets most easily found by Gaia have orbital periods measured in years; they therefore require years of observation before the Gaia astronomers can feel sure of their existence. The best estimates now suggest that in its first five years of observation, Gaia will find about 20,000 planets with Jupiter-like masses, moving in orbits somewhat larger than Earth's, and 70,000 of these planets if Gaia observes for 10 years. Several dozen should have orbits that undergo transits, providing further information along the lines described in Chapter 5. A small percentage of these planets will be found around M stars, typically at distances of a few hundred light years from the solar system, because the low luminosities of these stars makes them more difficult to observe than sunlike stars at the same distances.

By providing a survey of the numbers of Jupiter-like planets in our corner of the Milky Way, the Gaia results will furnish an accurate estimate of the likelihood that a star possesses a Jupiter-like planet, and how that likelihood depends on the star's intrinsic luminosity and its percentage of elements heavier than hydrogen and helium. These results will significantly improve our understanding of how, and how often, planetary systems that include at least one large planet have formed along with their stars.

In 2018, the European Space Agency announced plans for a mission to study exoplanet atmospheres by observing how they change the spectra of light from their stars during transits (see Chapter 5). Planned for launch in 2028, the Ariel (Atmospheric Remote-sensing Infrared Exoplanet Large-survey) spacecraft will study hundreds of exoplanets, most of them "hot Jupiters," in visible light and infrared, seeking information about their atmospheric composition that can reveal how and where the planets formed.

When Stars Shake: Asteroseismology Can Reveal Stellar Ages

The vast majority of stars have an internal structure that features an outer "convection zone," a region below the stars' surfaces within which currents carry hotter gas upward and cooler gas downward. These convection currents resemble what happens within a pan of water close to boiling. On the sun, the tops of the convection currents appear as "solar granulation," the apparent boiling of the solar surface. Astronomers, who have assigned the name "asteroseismology" to their studies of the motions of the outermost parts of a star, have naturally sought to apply this technique to stars hundreds of thousands or millions of times more distant than the sun. At these distances, the stars' overall convection currents appear either as modest fluctuations in the stars' total brightnesses, or, more directly, as slight, repetitive changes in the stars' radial velocities as their surfaces burble and boil.

Because the brightness variations amount to a few parts per million, and the changes in radial velocities are a modest fraction of 1 meter per second, detecting either of these phenomena poses a continuing challenge. The payoff, however, can be considerable: The details of asteroseismological observations provide astronomers with the *ages* of stars. The steady conversion of hydrogen into helium within a star's core slowly increases the density of the core and decreases the density in the star's outer layers. These changes affect the details of how the star shakes, and astronomers can detect the changes from careful asteroseismological observation. In the later stages of stellar evolution, as a star's outer layers swell enormously to create a red giant star, asteroseismology becomes even more useful in probing the details of what happens deep inside a star.[4]

The application of asteroseismological techniques to the estimation of stellar ages now remains in its infancy. The K2 mission will notably increase our ability to derive ages from these observations, and the TESS spacecraft should observe half a million stars to see how they quake.

PLATO: Kepler's Successor for the 2020s

The original Kepler mission and its follow-on K2 effort (discussed in Chapter 5) have shown how much astronomers can learn about exoplanets by carefully measuring their effects when they pass across the line of sight to their stars. The Kepler of the future, to be constructed and launched by the European Space Agency in 2026, bears the name of PLATO, a salute to the great Greek philosopher and an acronym for "PLAnetary Transits and Observations of stars." Although PLATO will live up to the final letter of its name by studying stellar oscillations—which can provide important facts about the interiors of stars and help to refine the results of transit observations—its primary goal touches a deep and persistent human interest: the search for habitable worlds similar to our own throughout the Milky Way. PLATO will be equipped with four groups of six cameras each, providing a wide-angle view of about $\frac{1}{20}$ of the total sky. This will allow it to monitor the brightnesses of hundreds of thousands of stars and search for the transits of Earthlike and super-Earth exoplanets that orbit within the stars' habitable zones.[5]

WFIRST: A Space Telescope for the Mid-2020s?

The future of finding exoplanets with astronomical microlensing lies with NASA's Wide Field InfraRed Survey Telescope, or WFIRST. This project embodies creative recycling on a grand scale, as the heart of the WFIRST telescope derives from a spy satellite that was never launched into space. The design of this satellite, one of a number commissioned by the United States National Reconnaissance Office (an organization not given to publicity, which describes its spacecraft creators only as "private contractors"), closely resembles that of the Hubble Space Telescope that Lockheed Martin Space Systems built four decades ago. WFIRST's 2.4-meter mirror, the same size as the Hubble Telescope's, captures and focuses infrared radiation for analysis by specialized detectors.[6] Although WFIRST will pursue many observational programs, including studies of the expanding universe, the nature of dark energy, and the direct observation of exoplanets (see Chapter 6), it will devote a significant part of its mission to the detection of exoplanets through observations of microlensing events. Freed from the constraints of the atmosphere, WFIRST, now scheduled for launch in the mid-2020s, should be able to detect planets with masses as small as one-tenth of the Earth's.

American Astronomers' Decadal Surveys and the Choice of Projects

The rise of WFIRST provides one of the more appealing examples of scientific cooperation. Every 10 years, since the 1960s, the United States astronomical community has helped to create a "decadal survey," under the aegis of the National Academies of Science, Engineering and Medicine, that reacts according to the strength of the community's desires in assessing proposed projects. The decadal surveys' recommended projects, usually ranked separately within two or three separate cost categories, emerge with the astronomical community's full weight behind them.

This consensus emerges only after strenuous argument. After a panel of experts deals with each subfield of astrophysics, the panels' recommendations move toward integration and final ranking. As one may easily imagine, tempers may flare over the selection of panel members, as well as during discussions and final deliberations; these issues include arguments over whether being too deeply involved in a project, even before a survey began, may call for disqualification. The survey's results, which often reflect the intense scientific and emotional involvement of its participants, have a serious impact on the congressional and other committees that consider providing funds for the decadal survey's list of future projects. For example, when WFIRST appeared at the top of the list that emerged from the 2010 survey, its chances of becoming a reality bounded upward.[7] Some would say that this result occurred because of the "Mars Rover effect"—the need to create a vehicle with something for everyone potentially involved in the project's outcome—and that this effect risks diluting the scientific results. WFIRST's backers regard this gamble as entirely worthwhile.

How to Overcome Diffraction
to Observe Exoplanets

Although WFIRST will have its own coronagraphic mask, at a cost estimated at $500 million, the spacecraft could eventually employ a markedly different approach to blocking the light from a star so that its planets become visible. Placement of a coronagraphic mask in a telescope's focal plane creates a series of problems, the most notable of which resides in the "diffraction" that the mask induces. Diffraction refers to the behavior of waves that encounter an obstacle and bend around it to produce a diffraction pattern of light that superimposes itself on what would be the image if no blockage occurred. This phenomenon often appears in high-quality images of stars, which acquire spikes of light, reminiscent of artistic depictions of a bright star, that arise from diffraction around the struts that hold the secondary mirror of a reflecting telescope.

The effects of diffraction limit any telescope's theoretical best ability to resolve fine detail in the objects that it observes. A large telescope's main mirror, for example, attempts to focus light from each point on a source onto a single point of its detectors—except that such a perfect point will appear only for a mirror with an infinite diameter. We can think of the different parts of a mirror as each being a separate surface, reflecting light waves in a particular direction, to form an image that combines the contributions from the multitude of individual directions. Larger mirrors will come closer to the ideal of producing a pointlike image, but they will inevitably fall short. The theoretical best outcome that the diffraction effect imposes on an instrument depends on the ratio of the wavelength of the observed radiation to the diameter of the lens or mirror used to focus the radiation toward a single point. An optical system that achieves this result is said to be "diffraction limited."

Because our atmosphere typically distorts light waves passing through it, even a telescope capable of reaching this diffraction limit rarely achieves this highest achievable resolving power. In space, however, the diffraction limit presents the natural goal of any telescope's design, and the desire to eliminate other forms of diffraction, such as those arising from anything placed in the path of the radiation being collected, gains strength. The finest way to achieve the telescope's diffraction limit allows nothing—such as a coronagraphic mask that creates diffraction patterns of its own—to interfere with radiation reaching the telescope. But how, then, can we hope to block the light from a star in order to observe its planets nearby?

The solution: Don't try to block the star's light *inside* the telescope! Instead, avoid the complex optics of that situation with a system that hides the star's light long before it reaches the optical system, so that the telescope has no chance to see the star. We can achieve this goal by creating a "starshade" far from the telescope and carefully positioning it so that the telescope effectively "sees" only blackness where the star would be without the starshade in place. The starshade concept originated in 1962 with the insights of Lyman Spitzer, the man who envisioned a telescope in space.[8] Seventy-some years later, technological progress may bring this idea to its practical application. Why the delay? When we address the optical facts of life, we find that the starshade should measure about 35 meters across and maintain a distance of approximately 40,000 kilometers from the telescope![9]

What was that? Place a starshade the size of a baseball diamond away from the spacecraft at a distance three times the Earth's diameter, and keep it orbiting in formation with the spacecraft with an accuracy sufficient to block the light from the star, but not from its planets? Sara Seager, the principal advocate of the starshade concept and a prominent exoplanet hunter who dreams (as so many do) of finding Earth's twin, left Canada to become a U.S.-trained astrophysicist and planetary scientist at MIT. In 2016, the *New York Times Magazine* described her subsequent personal history and career, full of tragedy and triumph, which led to her becoming a key player in the planning of future exoplanet-hunting missions.[10] She became the leader of a 2015 NASA study that proposed to equip the WFIRST telescope with a starshade of considerable scope. If all goes well after WFIRST has achieved successful operation in the mid-2020s, a starshade could be launched a few years later, ready to achieve what Seager calls a "starshade rendezvous" at the L2 orbital point, the location most favorable to the operation of spaceborne observatories (see note 1 to this chapter).[11]

Because M dwarf stars shine so much more faintly than sunlike stars do, their planets would be prime targets for WFIRST and its starshade. Should this plan become reality, Seager could continue to work toward the fulfillment of her greater dream: a still larger space telescope and starshade capable of direct observations of Earth-sized planets orbiting within the habitable zones that surround sunlike stars. These observations would include spectroscopic analysis to search for biosignatures that signal the possible existence of life (see Chapter 12).

Even as we may choose to assume that technological marvels can create, launch, deploy, and properly maintain a 35-meter starshade tens of thousands of kilometers from a spaceborne telescope, we must confront an additional difficulty, which arises from the need to direct the telescope and its starshade from one planetary system toward another. These shifts in the direction of observation will typically involve moving the starshade by tens of thousands of kilometers, an action that requires not only time but also fuel for the motors that propel the starshade and also, with considerably smaller energy consumption, maintain the starshade's proper position while the telescope studies a particular star. Current estimates of the fuel that a starshade system would carry (based, of course, on the cost of the mission), could limit the telescope's observational capability to a couple of dozen stars, but even with this reduced capability the telescope would still be sufficient to find a variety of planets in their habitable zones.

Visions of the More Distant Future: HabEx and LUVOIR

Looking beyond the already mind-boggling concept of equipping WFIRST with a 35-meter starshade out to the farther future of hunting for exoplanets, astronomers in the United States and abroad have proposed two great projects, both of which incorporate imaginative innovations as they (potentially) compete for always-limited funds.

These two spaceborne observatories carry the acronymic names of LUVOIR and HabEx. LUVOIR, the Large UV / Optical / InfraRed Surveyor, represents a mighty step beyond JWST, the current great successor to the Hubble Space Telescope. Much as JWST's 6.5-meter-wide main mirror far surpasses the 2.4-meter diameters of the mirrors of the Hubble Telescope and its physically similar cousin, WFIRST, LUVOIR's planners envision a much larger mirror, 9–16 meters in diameter, to give their telescope, even at low end of its possible size range, a far greater light-collecting ability. LUVOIR would also observe the cosmos over a significantly wider range of wavelengths and frequencies than those available to its predecessors. Deep in space, equipped with a starshade or a coronagraphic mask (or both), and orbiting the sun at a position far from Earth, LUVOIR could obtain direct images of exoplanets as small as the Earth, and then study these planets spectroscopically for atmospheric signs of life. LUVOIR would also study much more distant objects than these exoplanets, including the most distant galaxies that provide key clues to galaxy formation and the future of the universe's expansion, but the detection and analysis of exoplanets would rank close to the top among its goals.[12]

HabEx, the Habitable Exoplanet Imaging Mission, has a name that signals its planners' emphasis on the search for exoplanets capable of supporting life, though this spaceborne observatory would also perform valuable service in examining other objects throughout the universe. In contrast to LUVOIR, HabEx would be specifically designed to optimize its ability to study exoplanets. Its mirror would be somewhat smaller than LUVOIR's, and the telescope would have an "off-axis" design to avoid placing anything in the direct path to the mirror; instead, the mirror would reflect light a tiny bit sideways, so that the secondary mirror that directs the beam toward the instruments would not produce diffraction from that component. Whether HabEx would have a coronagraphic mask or a starshade remains a point of debate among its proponents. HabEx's projected cost would be much lower than LUVOIR's because of its comparatively limited capabilities. By studying the 20 or so planetary systems closest to the sun, HabEx would be able to obtain direct images and make spectroscopic observations of only the larger members of these systems, and it would remain unable to discern Earth-sized planets that the more ambitious and expensive LUVOIR could find and examine.[13]

Future Earth-Based Giant Telescopes

Back on Earth, three groups of astronomers stand ready to create a new era in ground-based observations of the cosmos during the next decade. Their three new telescopes, known by their acronyms as the EELT, the GMT, and the TMT, all draw on the concept that allows giant mirrors to maintain their shape within the accuracy (one ten-millionth of a meter) required for a telescope to yield accurate images. As a telescope turns to move from one object to another, or to follow an object across the sky, changing gravitational flexes tend to deform its primary mirror, rendering it unable to focus light with an accuracy sufficient to justify its size and cost.

Seventy years ago, the Palomar Observatory's five-meter giant reached what for many decades remained the maximum possible size for a single glass mirror in a telescope on Earth. Eventually, new technology allowed the creation of much lighter and thinner mirrors with diameters as large as eight meters. The current record holders are the twin 8.4-meter mirrors of the Large Binocular Telescope on Mount Graham in Arizona. The telescope's architect, Roger Angel, pioneered the creation of "honeycomb" mirrors that reduce the mirrors' weight by factors of 4 or 5 over conventional designs. A crucial step toward the design of even larger mirrors began with two California telescope designers, Jerry Nelson and Terry Mast, who demonstrated that newly available technology allowed the creation of segmented mirrors, composed of individual mirrors whose edges could be continually realigned by activating motors behind them. These motors, operating once or twice a second, would move the segments through tiny distances, allowing the mirror that they create to maintain a near-perfect surface. The phrase "active optics" describes this approach to creating an astronomically useful mirror much larger than any single disk of material could provide.[14]

Mast and Nelson's idea, now commonplace for the design of all large telescopes, achieved reality in the twin Keck Telescopes on Mauna Kea, each of which has a primary reflector made of 36 separate hexagonal mirrors, each measuring 1.8 meters across, that together form a 10-meter mirror, with twice the diameter and four times the light-gathering power of the Palomar giant. In 2009, a consortium of European countries followed the Keck plan to inaugurate the slightly larger Gran Telescopio Canarias, whose 10.5-meter mirror also consists of 36 hexagonal segments. This telescope, sited at the Roque de los Muchachos Observatory on the island of La Palma, enjoys the excellent weather that prevails in the Canary Islands, though its altitude of 2,267 meters leaves it below significantly more of the atmosphere than the Keck Telescopes at their 4,205-meter altitude.[15]

A third large segmented-mirror instrument, the Southern African Large Telescope (SALT), follows a slightly different design, but it also creates its primary mirror from hexagonal segments; in this case, 91 segments, each measuring 1 meter across, together span 9.8×11.1 meters, thus creating the largest telescope now operating in the Southern Hemisphere. Sited within a nature reserve northeast of Cape Town at the comparatively low altitude of 1,798 meters, the telescope began operation in 2005 after drawing on funds from South Africa as well as the United States, Germany, Poland, the United Kingdom, and New Zealand.[16] All of these giant segmented-mirror telescopes employ not only active optics but also the adaptive optics described in Chapter 6, which compensate, on timescales measured in hundredths of a second, for the distortions that arise from light waves' passage through the atmosphere.

The success of telescopes embodying segmented architecture naturally inspired astronomers to look toward even greater instruments that would employ the segmented-mirror principle. The estimated costs of $1 billion–$1.6 billion for each of these future behemoths require that funding for their planning, construction, and operation must come from consortia of multiple institutions or national agencies (or both).

Of the three planned telescopes that aim to surpass any now in operation, two currently hope to achieve "first light" by the mid-2020s. The first of this duo, the Giant Magellan Telescope (GMT), began construction in 2015 at an altitude of 2,516 meters at the Las Campanas Observatory near La Serena in north-central Chile. With funding from universities and research institutions in the United States, Australia, Brazil, and South Korea, the GMT will employ seven large mirrors, each of them 8.4 meters across (thus matching in size of the mirrors of the Large Binocular Telescope in Arizona); six of those mirrors will surround the central seventh. The asymmetric shapes of the mirrors required to make this design work leave them especially difficult to fabricate. Collectively, the seven mirrors will provide a reflecting surface 25.4 meters in diameter. Astronomers now plan to begin operation with four of these mirrors in place in 2021, and they are working toward having the full sevenfold complement in operation a few years later.[17]

The second of the extremely large telescope projects, the European Extremely Large Telescope, or EELT, certainly deserves its name. The telescope's funding comes from the European Southern Observatory, the organization of 16 countries that has already built and operated several observatories in northern Chile; its members contribute funds to astronomical projects in proportion to the size of each country's gross domestic product. ESO's current crown jewels include the Atacama Large Millimeter Array (see Chapter 11), which it created in collaboration with the United States and Japan, and the four 8.2-meter telescopes of the Very Large Telescope array at the Paranal Observatory that, when used in tandem, form the world's largest visible-light telescope. The EELT aims to put these (and the GMT) in the shade by employing 798 hexagonal mirrors, each of them measuring 1.8 meters across, to create a mirror 39.3 meters in diameter, with a light-gathering ability nearly 16 times of either of the Keck Telescope's twins. Construction of the giant telescope, situated atop the 3,046-meter summit of Cerro Armazones beneath the amazingly clear skies of the Atacama Desert in northern Chile, started in 2017. The EELT's builders hope to begin operation of their great instrument in 2024.[18]

The third of these mammoth undertakings now lies enmeshed within a familiar American problem: local politics. Known as the

Thirty Meter Telescope, or TMT, this telescope ranked highest among the ground-based projects that were recommended in the decadal astronomy survey released in 2001. The first decade of this millennium brought the telescope's design close to completion, with funding commitments from the Gordon and Betty Moore Foundation, Caltech, the University of California, Japan, and eventually India, Canada, and China. With a design and a construction schedule similar to the EELT's, the TMT seemed on track to join their rise as the world's third extremely large telescope, one that would combine 492 hexagonal mirrors, each measuring 1.44 meters from corner to corner, to form a primary mirror 30 meters across.[19] A detailed survey of the best possible sites for the TMT led to the 2009 selection of the summit of Mauna Kea, partly to provide a Northern-Hemisphere location that would complement the Chilean telescopes, but mainly because Mauna Kea offers superbly clear skies, free from most of the water vapor that resides in the lower atmosphere and absorbs infrared radiation.

The Mauna Kea Observatory already hosts the sites for 11 astronomical instruments—and that's a problem. Native Hawaiian culture regards the summit of Mauna Kea as sacred ground, which the Historic Preservation Act of 1986 has recognized. The compromise that led to the creation of the observatory during the 1960s specified that all telescopes would remain within an "astronomy precinct" covering less than a square mile on the volcano's summit. In Hawaii, groups opposed to the new telescope, mainly native Hawaiians and those who feared the environmental impact of successively larger construction projects, originally found themselves with less influence than those who welcomed opportunities for construction and scientific investigation.

As time passed, however, the opposition gained strength and legal sophistication as they rejected a compromise proposed by Hawaii's governor that would have created a cultural council to oversee the summit and require the removal of some of the existing telescopes. At the end of 2015, the Hawaii Supreme Court halted the TMT construction (already in progress) by revoking its construction permit. Although a judge of the Hawaii land court ruled in 2017 that the permit should be approved, and the Hawaii Board of Land and Natural Resources agreed with this decision, continuing legal and political fights may eventually delay the Mauna Kea option to the point that the TMT's supporters will abandon their plans for this site and relocate the project to the Canary Islands. In any case, construction of the TMT should almost certainly begin before this decade draws to a close, with completion expected a decade or so later.[20]

The era of extremely large telescopes will bring a host of opportunities for improving our understanding of cosmic objects and events. Exoplanet studies rank near the top of the list of anticipated gains, along with observations of the light from the faintest, most distant galaxies, seen as they formed more than 10 billion years ago. For the first time, ground-based telescopes will make detailed spectroscopic observations of faraway worlds in the Milky Way and employ instruments far more massive and advanced than those any spacecraft can carry. These visible-light and infrared spectroscopes can analyze the atmospheres of planets orbiting within the habitable zones of their stars, and thus find—if they exist—molecules of water vapor, carbon dioxide, oxygen, and other compounds. In other words, these new telescopes will be able to determine whether exoplanets situated for life in fact show suitability for life, and perhaps even the presence of life itself.

For a Real Picture, Use Interferometry

Giant telescopes in space, equipped with the finest star shades and coronagraphic masks, may serve well to provide direct images of planets orbiting comparatively nearby stars. However, by itself, even the largest telescope that we may reasonably envision cannot satisfy our fundamental curiosity about the exoplanets that most closely resemble the Earth in size and mass. This stubborn fact arises from the laws of optics, which describe any optical system's theoretical best ability to resolve fine detail. That ability depends on two factors: the size of the lens or mirror used to make an observation, and the wavelength of the radiation under observation. Longer wavelengths provide coarser views of reality, lowering our ability to observe fine detail. Larger apertures—the diameter of the lens or mirror brought to bear upon a source of radiation—increase the "resolving power" of the instrument employed in our efforts.

This resolving power depends on a single ratio: the wavelength used in our studies, divided by the aperture of the system. Because both of these numbers are lengths, their ratio becomes a single number. This number provides the theoretical best resolution of detail, so long as we measure angles in what scientists consider the natural unit of "radians," each of which equals $360/(2\pi) = 57.296$ degrees. (For complete accuracy, we must multiply the number that we derive for the best angular resolution by a factor sufficiently close to 1 for us to ignore it in this discussion.) If, for example, astronomers use a reflecting telescope whose diameter equals 40 meters to observe a source of visible light whose wavelength equals 400 nanometers (400 one-billionths of a meter), this analysis shows that they could, in theory and under perfect observing conditions, distinguish any two equally bright sources whose angular separation exceeds 2.1 one-thousandths of a second of arc.[21]

In theory, then, we can observe any details on a planetary surface that subtend angles at least as large as this value. What amazing results could flow from placing a 40-meter telescope in space! On the ground, the atmosphere limits the resolution of such a telescope—the same size as the EELT—to about 0.05 second of arc, which is 25 times worse than the theoretical best resolution achievable in the atmosphere's absence.

Suppose that we observe an exoplanet at a distance of 10 parsecs. One parsec, which corresponds to 3.26 light years, specifies the distance of a star for which the Earth's yearly motion around the sun would cause an apparent back-and-forth shift by 1 second of arc in the star's apparent position. What features on the surface of such an exoplanet could we hope to resolve with a 40-meter telescope in space? Trigonometry shows that for a planet 10 parsecs (32.6 light years) from us, our theoretical best minimum corresponds to a distance of about 3 million kilometers! If we want to observe regions that span not 3 million kilometers but, for example, 300 kilometers, we would require a telescope with a mirror not 40 meters across but 10,000 times wider: 300,000 meters, or 300 kilometers!

These numbers pose a serious challenge to astronomers who hope to examine the surface features of exoplanets. Human ingenuity, however, has found a solution—in theory. We don't actually need to construct a telescope that measures 300 kilometers wide for our purposes. All telescopes perform a twofold function. First, they gather light, and because larger telescopes gather more light, they can detect fainter sources. Second, they provide the resolving power that we have been considering. If we observe comparatively bright objects, for which our light-gathering needs prove comparatively modest, then we can create an array of smaller telescopes and link their observations simultaneously to create the resolving power of a single telescope as large as the array.

Astronomers call such an array of linked telescopes an "interferometer" because the array's ability to combine all the telescopes' observations depends on the phenomenon of interference—the fact that when light or other types of waves encounter one another, they can either combine with each other or cancel one another, depending on whether the wave crests coincide and are "in phase," or, in the opposite case, cancel because they are "out of phase."

Happily, we have already encountered enough physics for a single chapter, we and need not dwell long on interference. We could study this phenomenon for sound waves in person through the purchase of expensive headphones that use interference to cancel the ambient noise that would otherwise reach our ears. Instead, we may note that astronomers have long put the principles of interference into practice—not in space, nor so much with light waves, but with radio waves, whose much longer wavelengths make them easier to manipulate in order to achieve the desired results. Starting in the 1970s, astronomers created large arrays of radio telescopes, most notably the Chilean ALMA array designed to observe shorter-wavelength radio waves (see Chapter 11). To observe conventional radio waves, astronomers can use the pioneering Very Large Array on the high plains west of Socorro, New Mexico, which consists of 27 antennas, each measuring 25 meters in diameter, that spread along the three 21-kilometer-long arms of a Y-shaped array. The antennas move on railroad tracks to assume various configurations (one of the tracks crosses U.S. Route 60 at grade, causing occasional motorist delays), and, at their maximum extent, and at the shortest radio wavelengths used for the telescope array's observations, the antennas can achieve a resolving power of 0.02 arc seconds.[22] Thus the Very Large Array can achieve about one-tenth of the resolving power of a 40-meter telescope in space that observes light with a wavelength of 400 nanometers. The difference in the "diameter" of the two observing systems—40 meters versus 40 kilometers—reminds us of the difference in wavelength: The radio waves in question have wavelengths about 10,000 times longer than 400 nanometers, so the diameter of the radio array would have to be about 10,000 times larger in order to mimic the ability of a telescope observing in visible light. Instead, the array is "only" about 1,000 times wider than a 40-meter telescope mirror. Plans are now underway to create the Next Generation Very Large Array, which would be 10 times larger than the current configuration and therefore provide a 10-fold increase in the array's ability to resolve fine detail in mapping the sources of radio waves.[23]

What stands in the way of creating a spaceborne interferometer system that spans hundreds of kilometers? The chief obstacles are money and technological issues, neither of which would pose much of a problem for a society with untold resources and plenty of time. Astronomers have been planning spaceborne interferometer systems capable of obtaining comparatively high-resolution images of planetary systems for many decades, long before they had any specific planets to dream of targeting. Among the leading projects, none of which has current support, we should cite NASA's Terrestrial Planet Finder, or TPF, and its European counterpart, the Darwin mission of the European Space Agency (ESA), whose name for once was not an acronym but an homage to the great naturalist.[24] Both of these missions envisioned the creation of four to nine reflecting telescopes of comparatively modest size, with mirrors three or four meters across, that would fly in formation and beam their observations to a satellite in the formation's center, where interferometry would create a single, most excellent image. A far smaller interferometric observatory, NASA's Space Interferometry Mission (SIM), would have mounted two telescopes at opposite ends of a six-meter beam. This modest size would have allowed SIM to make highly accurate astrometric measurements that could reveal Earth-sized planets around the closer stars by measuring the stars' deviations from straight-line motion through space (see Chapter 3).

NASA planned to construct, launch, and operate SIM as a way to demonstrate the feasibility of spaceborne interferometry before introducing the far more advanced TPF. In the early years of the new millennium, this seemed a reasonable plan, and a friendly, if expensive rivalry between NASA and ESA for interferometric superiority seemed in the offing. As things developed, however, what was offed was the funding for all of these projects.[25]

The demise of the TPF project has naturally diminished the importance of a hard-fought battle between the two teams of scientists involved in determining its design. This struggle pitted visible-light astronomers against those favoring observations in the infrared. Each of these domains offers both advantages and drawbacks in the search for exoplanets and the possibilities of examining them in detail. As we have seen, in visible light stars typically outshine their planets by a factor of a billion or so, whereas in the infrared this number diminishes a thousandfold, to about a million. On the other hand, visible light offers much richer possibilities for examining a planet without overwhelming interference from the star nearby.

Optical interferometry does exist and has demonstrated its usefulness ever since Albert Michelson, the first American to win the Nobel prize in physics, employed an interferometer at the 100-inch reflecting telescope at the Mount Wilson Observatory in California to make the first measurement of a star's diameter—in this case, the red supergiant Betelgeuse in Orion's shoulder. Nearly a century later, astronomers have created optical interferometers at the Palomar Observatory in California and the Mauna Kea Observatory in Hawaii. At the latter observatory, the twin 10-meter Keck telescopes functioned, on occasion, as an interferometer system from 2003 to 2012. Astronomers using this system aimed primarily at measuring the sizes of the disks of dust-rich gas that surround many young stars, and they were fully aware that an Earth-borne interferometer consisting of two mirrors, separated by 85 meters, could hardly begin to furnish images of exoplanets.

The Far, Sweet Future:
Coronagraphy plus Gravitational Lensing

Among those unafraid of dreaming big, Slava Turyshev of the Jet Propulsion Laboratory deserves a salute. Turyshev, one of the few astronomers with a PhD (from Moscow State University) and an MBA (from UCLA), imagines a future in which we can obtain detailed images of exoplanets from an unsuspected gravitational lens: our sun. Working in collaboration with his colleagues at JPL, Turyshev has developed the mathematical analysis behind a plan to use the sun's gravitational force to produce an "Einstein ring" (described in Chapter 7) that appears when a massive body stands directly between an observer and a much more distant object. The bending of space deflects light from the distant object in all directions around the intervening body, and this distorted ring of light provides an image of one particular spot on that body's surface. By slightly moving the observer's position, we can record the Einstein ring from each point on the body's surface and can then reconstruct, pixel by pixel, an actual multipixel image of the faraway object.[26]

Some difficulties arise in the actual implementation of this project. First, the spot where the sun's gravitational force creates the Einstein rings to be used in this effort lies about 550 AU from the sun. We would therefore require a spacecraft capable of reaching this distance on a timescale sufficiently short to motivate funding for the project. Although our highest-velocity spacecraft currently travel at a bit more than 3 AU per year, technological improvements might before long increase these speeds by a factor of 5 or so, thereby reducing the journey time to a couple of decades. Because the spacecraft must have the ability to come to a halt and then maintain its position with respect to the sun accurately, we can hardly imagine the use of anything like the low-mass, laser-accelerated interstellar voyagers described in the next chapter. Instead, more conventional spacecraft propulsion would have to furnish the means of reaching the immense distances required for our purposes.

Second, because the gravitational-lens method requires the inspection of images formed by the passage of light rays close by the sun, the observational task must include a means of blocking the sun's light. This task might be achieved with a "conventional" coronagraph, which places a mask in the optical system, directly in front of the sun as we observe its immediate surroundings. Alternatively, we could create a separate sunshade that blocks the sun's light before it reaches the optical system that observes the Einstein rings. As we discussed earlier in this chapter in the context of observing planets orbiting other stars, each approach has its advantages and drawbacks.

We may safely conclude that although Turyshev's proposal merits admiration and consideration, its execution lies well into our astronomical future. This conclusion gains strength when we consider that the placement of an Einstein-ring observatory at 550 AU from the sun works for just one planetary system—the one directly behind the sun, as seen by the spacecraft. We could, of course, create as many of these observatories as our determination and budget would allow, and some might say that if we did secure a detailed image of a single exoplanet's surface—especially one that seems particularly favorable to life—we would highly increase our motivation to continue along this path of the study of the most interesting exoplanets.

14

.

PROXIMA CALLS:
CAN WE VISIT?

n 2016, Proxima Centauri, the sun's closest neighbor star, provided a psychological boost to those who dream not merely of observing, but actually visiting planets that orbit other stars. Conditions on Proxima b, the Earth-sized planet in orbit around this low-luminosity M star, may or may not favor life on its surface. The best way to find out could involve direct visits to this faraway world.

That we may consider this possibility worthy of discussion does not, of course, assure us that it will soon become reality. Nor, we should emphasize, does it have anything to do with sending human visitors to extrasolar worlds. While imposing, at least temporarily, these important limits on our soaring imaginations, we may admire the fact that we do have a chance, well within some of our lifetimes, of sending probes to the closest stars and their planets and gleaning significant amounts of information from these exoplanet investigators.

The technological breakthrough that could enable the previous sentence to pass from possibility to reality resides in the concept of "nanoprobes," tiny, wafer-thin assemblages of data-gathering and broadcasting equipment that could be accelerated by laser beams to a sizable fraction—20 percent, perhaps—of the speed of light, which equals just about 300,000 kilometers per second. At 20 percent of the speed of light, the 4.22-light-year journey to Proxima Centauri would take only 21 years. If we add the light-travel time required for the probe's observational results to reach the Earth, then we face a mere quarter-century of elapsed time between the launch of the probe and the arrival of its data. Gone are the thoughts of automated probes that require centuries to reach the nearest stars, or the cinematic visions of astronauts spending thousands of years in suspended animation on the same sorts of journeys. Instead, we have (mentally!) broken the time-laden barrier and stand ready for the comparatively rapid exploration of nearby planetary systems.

Before we ease into an examination of how laser acceleration of interstellar probes might actually work, let's pause to admire the leap of distance that our minds have just performed. In contemplating the enormous distance scales that characterize the universe, two ratios involving the number 1 million help to keep the vast spacing of the cosmos in proper perspective. First, the distance to the three stars in the closest star system—Alpha Centauri A, B, and C (Proxima)—exceeds the distance to Mars by roughly a quarter of a million times, depending on the relative positions of Mars and Earth in their orbits.

Second, the nearest large galaxies beyond the Milky Way lie about a million times farther from us than Alpha Centauri (for the Andromeda galaxy, that ratio is only about half a million, but we are rounding off). Thus a trip to the Andromeda galaxy would span a distance somewhat less than a *trillion* times the distance to Mars. (For continuing accuracy, we should recognize that a trip to Mars will never proceed along the most direct and shortest route, but instead it must take advantage of the Earth's speed in orbit to follow a longer trajectory.)

Now that we are firmly aware of these enormous distance factors, we recognize that we can hardly expect to send probes to other galaxies, even if we manage the near-incredible (for now!) feat of sending a tiny spacecraft to the closest stars. Since 2015, NASA has supported research into the possibility of achieving this audacious goal—which, predictably, requires entirely new paradigms for the construction, launch, and operation of an interstellar spacecraft.

The driving force behind this potentially revolutionary approach to exoplanet observations can be found at the University of California, Santa Barbara, where Philip Lubin, a cosmologist who has been long engaged in balloon- and spaceborne studies of the cosmic background radiation, has turned his major attention to the issues raised by the human desire to travel to the nearest stars. The first order of business in the quest to fulfill this dream seeks a spacecraft system that can reach far greater speeds than those attainable with the chemically propelled rockets that carry our probes through the solar system. These explorers reached their maximum velocity to date—16.26 kilometers per second—in the New Horizons spacecraft that passed by Pluto in 2015.[1] Since Proxima Centauri lies just over 40 trillion kilometers away, a rocket traveling at this speed would require 2.46 trillion seconds, or 78,000 years, to complete a one-way journey. Lubin and his colleagues aim for a speed nearly 4,000 times greater than our current best—20 percent of the speed of light—which would reduce the travel time to just two decades.

To achieve this goal requires much smaller probes and a far mightier means of acceleration. Each of these interstellar travelers would have a mass of only a few grams, and each would be shot into space by a mighty array of lasers—50,000 or so—that would briefly require a power output greater than that consumed on a continuing basis by the state of California. Catching the combined laser beams onto its highly reflective surface, the probe would take about 20 seconds to reach one-fifth of the speed of light. After that amazing burst of acceleration, the probe would already be too distant from Earth for the lasers to focus their energy directly onto it, but the speed that it attained during these few seconds would allow the probe to coast for a couple of decades before completing its mission while passing by the nearest star and its planet.

Some serious technical challenges stand between this plan's concept and its fruition. The construction and operation of the lasers seems possible, not least with an admiring look at how far laser systems have improved since Charles Townes first envisioned their operation a long lifetime ago. These lasers would apply their energy to a probe, which would measure perhaps 1 meter in diameter but only a few millimeters in thickness and whose surface should reflect 99.9999 percent of the energy that strikes it, lest it burn from the incoming flux. This lightweight spacecraft would incorporate a camera, computer, and radio equipment capable of securing images and sending them back to Earth, as well as a modest propulsion system that could turn the probe onto its side during its journey, and then flip it into the best position for obtaining images and any other measurements from miniaturized instruments.

The requirement for the probe's sidewise turn arises from the presence of atoms, molecules, and dust grains of various sizes, relics from the great era of star formation in the Milky Way, billions of years ago, that are scattered through interstellar space. For a spacecraft that travels at 20 percent of the speed of light, each of these particles amounts to a tiny bullet, capable of striking the probe at just this velocity. Happily (more or less), larger particles are rarer than smaller ones: Interstellar space contains about one proton in every cubic centimeter, but you will find a dust grain perhaps a tenth of a micron across—millions of times larger than a proton—only if you sift through a million cubic meters of our interstellar neighborhood.

Those who support the idea of sending such nanoprobes to the closest exoplanets assert that these bullets do not seem to pose a killer problem, or, more accurately put, they suggest any actual probe-killing can be dealt with appropriately by launching more spacecraft. One of the possibly overlooked aspects of spacecraft launched by lasers toward the closer stars resides in the fact that we could construct and launch hundreds or thousands of them at comparatively low expense: A one-gram probe, no matter how miniaturized, would cost far less than the 500-kilogram New Horizons spacecraft, especially if we developed mass-production techniques for the one-gram probes' manufacture. In the longer term, we would learn from the successes and failures of the first generation of probes how to improve later generations. To those who doubt that a one-gram probe could perform as desired, Lubin points out that the entire electronic workings of a smartphone—the camera, the apps, the processors, and everything else—have a similar weight, protected by much more massive cases and read out on a comparatively large screen.[2]

What might these probes find once they reorient themselves into their most effective observing position and pass by Proxima Centauri and its planet? To travel a distance of 1 AU at 20 percent of the speed of light takes just over 40 minutes, which describes the interval that would allow the best views of the star–planet system, with special attention to Proxima Centauri b. Later generations of space probes could carry spectroscopic systems to analyze the surface and possible atmosphere of this or other planets in great detail, and thus to discern—potentially—the signatures of life-supporting gases, or even of life itself.

If we assign a quarter-century to the interval between the launch of these probes and the receipt of any information they might secure at their goal, and if assume that we might create these miniature investigators within a few decades, then we may reach a happy speculative result: By the end of the twenty-first century, we could be awaiting data from the second generation of interstellar probes, and we could be planning—if humanity survives to take the long view—farther-ranging spacecraft that would extend our investigations to the dozen nearest stars and their planetary systems.

For those who prefer a more mundane approach to laser-launched space probes, Lubin points out that the same sort of system could send packages to the moon in a few hours, and to Mars in a week or so. His team now inclines toward descriptors for these varied potential travelers that include "Starlight," "Moonlight," and (you guessed it) "Redlight."

Many who have heard of Lubin's plans associate them with the publicity boost that they received in April 2016, when the Russian-born technologist Yuri Milner announced his support for this laser-launched initiative. With an impressive board of advisors that included Stephen Hawking, Freeman Dyson, Martin Rees, former astronaut Pete Worden, and the Nobel-prizewinning physicists Stephen Chu and Saul Perlmutter, Milner's project drew immediate public attention for its projected happy outcome, its pledge of $100 million to support Lubin's goals, and its claim that success could occur within the next generation of space explorers. If this funding materializes, it could indeed give a significant boost to the interstellar dream that Lubin hopes to make reality. (Feeling the general constriction of NASA's long-term approach for planning and execution, Milner has also proposed a privately funded mission to Saturn's moon Enceladus in order to explore the possibilities of life in its subsurface ocean; see Chapter 12).[3]

Don't Forget Lasers for Communication!

Astronomers and others who speculate about the best ways to search for, and possibly to find, other technologically advanced civilizations often point toward laser beams as one of the best ways for a civilization to signal its existence toward a particular direction. The concept of our own low-mass probes accelerated by powerful lasers leads to an enjoyable speculation: We might happen to find ourselves within the trajectory of probes sent from another civilization, presumably (or do we presume too much?) because those probes are directed toward the solar system. In this situation, we would certainly have a better chance of detecting the laser beams that propel the probes rather than the probes themselves, especially if they have diameters less than a meter and do not pass close to Earth—the more so if they pass by at 20 percent or so of the speed of light.

Space Travel at Near-Light Speeds

Visionaries who dream of travel to the closest stars do not end their speculations with a modest vision of hordes of tiny probes being sent to Proxima Centauri at 20 percent of the speed of light. Instead, they look far beyond the confines of our present technology and ask, what about travel at speeds as close to light speed as we may imagine, with spacecraft that carry humans?

This question opens far greater possibilities than merely shortening the duration of future interstellar journeys. These missions would continue to require a few years for travel to the sun's very nearest neighbors, or tens or hundreds of years to make a useful survey of the few dozen closest planets that offer conditions apparently hospitable to life. Double these times to allow for the return journey, and the trips would last a lifetime or more.

But would it? Not if travel velocities rise so close to the speed of light that the effects of Einstein's special theory of relativity play an important role.[4] First published in 1905, Einstein's theory includes the effects of "time dilation," the slowing down of time that occurs for anyone who travels at speeds close to the speed of light. Time dilation offers a great advantage to future interstellar astronauts, allowing them to complete journeys that span dozens of light years while they age by only a few years. For example, an astronaut moving at 80 percent of the speed of light would age only 60 percent as rapidly as her twin left on Earth; at 95 percent of light speed, she would age 31 percent as rapidly, and at 99.5 percent of the speed of light, her aging would occur only 10 percent as rapidly. Thus a round trip covering a total of 100 light years would bring the astronaut back to Earth at an age 90 years younger than that of her twin who stayed behind!

But wait a minute—didn't Einstein say that all motion is relative? Why can't the twin on Earth say that she does the traveling? After all, she certainly moves with respect to the twin on the spaceship. What privileges one motion over another? Why doesn't each twin see the other as aging more slowly?

This "traveling twin paradox" has a resolution—and it's a good thing too, if we hope to continue to respect Albert Einstein as the genius that he was. The situation may seem symmetrical so far as the two twins are concerned, but on closer inspection, actual symmetry does not exist. The crucial difference resides in the fact that the traveling twin (as we may reasonably identify her) turns around in her journey. Those who find this statement insufficient may enjoy the next few paragraphs, which provide actual numbers in an example meant to demonstrate the facts that discriminate between the two twins.

Let us imagine a journey at 99.5 percent of the speed of light that covers the 45-light-year distance to 51 Pegasi b, the first exoplanet found by the radial-velocity method. Imagine further that both the traveling twin and her sister on Earth wear beacons that flash once per second. (This thought experiment arises from a suggestion by the physicist David Mermin, whose highly interesting presentation of Einstein's special theory of relativity appears in the Further Readings section of this book.) As the trip unfolds, each twin continuously monitors how rapidly flashes of light arrive from the other twin.

As the journey proceeds, two factors have paramount importance: the changing distance between the twins, and the slowing down of time predicted by relativity theory. At 99.5 percent of the speed of light, time slows down by a factor of 10, so each twin (rightly!) "sees" time unfolding 10 times more slowly for her sister as she (at first) travels away from her twin at near-light speed.

Take the traveling twin first. As she hastens on her way to 51 Pegasi, the slowing down of time that she perceives as her twin on Earth recedes from her at 99.5 percent of the speed of light would seem to cause the traveler to receive not one flash per second but instead one flash every 10 seconds—until we allow for the steadily increasing distance, which requires that each flash must travel farther to reach her from Earth. Between each received flash of light, the spaceship increases its distance from Earth by 9.95 light-seconds, so that light must travel for 9.95 seconds more to reach the traveling twin. Hence each successive flash of light reaches the traveling twin 19.95 seconds after the preceding one did, so long as the twin proceeds along her outward journey.

Meanwhile, back on Earth, the situation provides complete symmetry: The stay-at-home twin likewise receives flashes from the traveling twin at intervals of 19.95 seconds. If the traveling twin never changed direction, each twin would be entitled to say that time proceeds 10 times more slowly for the other twin. As Einstein pointed out, this creates no paradox, because the twins have no way of comparing their situations at any particular time. Only when and if the twins reunite can they determine who has aged more.

Now consider what happens when the twin reverses her direction of travel. (For simplicity, we ignore any time spent researching the possibilities of life on 51 Pegasi b.) As soon as this happens, the traveling twin receives flashes from Earth 20 times per second. Once again, this result depends on two effects: The slowing down of time would make the flashes appear to be emitted at intervals of 10 seconds, but in this interval, the traveling twin moves 9.95 light-seconds closer to Earth. The time between received flashes therefore equals 10 seconds minus 9.95 seconds, or 0.05 second. Crucially, the two different intervals of arrival times—once every 20 seconds on the outbound journey, 20 times per second on the inbound one—occur for equal amounts of time, as observed by the traveling twin.

Now consider what the Earthbound twin will record. She, too, will first record one flash every 20 seconds, and after that 20 flashes per second, but the two flashy rhythms will appear over highly unequal intervals of time. The Earthbound twin will see the changeover only when her twin has almost returned home. Long after the traveling twin has changed direction, her once-every-20-seconds flashes from her outbound journey will still be reaching the Earth at those intervals of time. The flash from her most distant location, 45 light years away, will arrive only two and a half months before the traveling twin herself comes home!

A detailed count of the flashes seen by each twin will verify the predictions of Einstein's relativity theory. To cover the round-trip distance of 90 light years at 99.5 percent of the speed of light requires the traveling twin to spend 90.45 years, as measured on Earth. But this twin will age by only 10 percent of this amount, or just over 9 years. "Relativistic travel"—a journey at speeds close to the speed of light—offers real anti-aging possibilities.

If none of this discussion has proven the point—or indeed anything beyond the fact that numbers such as those presented here always find their friends and enemies—readers may draw reassurance from the fact that the slowing down of time at near-light velocities has been repeatedly verified here on Earth. One method has involved placing amazingly accurate clocks in airplanes and showing that they record time a tiny bit more slowly than stationary clocks, even though their speeds reach only a tiny fraction of the speed of light. Other experiments do involve near-light velocities, attained both by the "cosmic ray" particles that continuously rain onto the Earth, and by others accelerated in elaborate machines designed for the purpose. Because physicists know the rates at which some of these particles "decay" into other particle types, they can verify that these decays proceed more slowly for the fast-moving particles, in a manner that matches Einstein's predictions.

Special relativity theory therefore offers a royal road to interstellar travel, whether of exploration or colonization. Instead of imagining journeys that take decades, centuries, or millennia, with astronaut crews dedicated to their task by reproducing successive generations of astronauts as they travel through space, we need only imagine travel at near-light speeds that can carry humans over distances of tens, hundreds, or thousands of light years within a single lifetime. The only drawback, if one perceives it as such, lies in the fact that generations would pass on Earth while the high-speed journey unfolds. The news from the travelers would reach not the society that launched them but their *descendants,* who might be all the more (or less) grateful for their ancestors' investment in the human desire to explore the cosmos.

Human Colonization of Habitable Planets

The prospect of astronauts embarking upon interstellar journeys that cover dozens or hundreds of light years, but age them by only a few years, leads naturally to the possibility of sending colonists to some of the closest habitable planets. One may reasonably conclude that humans will obtain the capacity to secure detailed images of these planets well before they achieve the ability (or perhaps the desire) to send human crews toward them at speeds close to the speed of light. These images, along with even better measurements made by lightweight probes to speed by these planets, would allow careful selection of the most promising sites for future human habitation. When we look back through the technological advances of the twentieth and early twenty-first centuries, we may conclude that achieving these goals by the end of this century, or the middle of the next, does not violate rational possibility.

The twenty-first century, unlike the centuries that came before it, offers the real possibility that human colonization of other worlds could soon begin on planets much, much closer to us—our solar-system neighbors—rather than on the exoplanets explored in this book. In contemplating these prospects, we encounter a threshold issue: Does a moral question arise from our plans to land on other worlds, even those with no signs of life or hints that life might once have existed, and to work our will upon them?

Consider, for example, our own moon, or the equally lifeless, though smaller, asteroids that orbit the sun, primarily beyond the orbit of Mars. Many of these objects contain copious amounts of valuable minerals and other compounds. The moon's surface, for instance, contains large amounts of helium-3, a comparatively rare isotope of helium that plays a vital role in the nuclear-fusion reactors that might someday provide enormous amounts of clean power on Earth or for space colonies. In the more shaded portions of the lunar surface, the concentration of helium-3 may equal as much as 50 parts per billion, a tiny fraction but one many times higher than the corresponding value for helium-3 on Earth. A few hundred million tons of lunar soil could yield 20 tons of helium-3. Does this benefit outweigh any problem created by a lunar mining operation on the scale needed to achieve this result?

If reworking the moon seems a bit much, what about mining on one of the small number of asteroids that consist mainly of metals, as opposed to mining on one of the vast asteroidal majority, which are mostly made of rocks? Unlike the Earth, within which much of the heavier-element contribution has sunk inward to join the core, a metal-rich asteroid, too small to possess a structure differentiated into an inner core and an outer mantle, would provide easy access to its heavy elements. The long list of desirables includes iron, nickel, magnesium, manganese, cobalt, silver, and gold, as well as rarer elements such as molybdenum, osmium, palladium, platinum, tungsten, ruthenium, rhenium, and even rhodium, all of which now have important applications in modern technological life.

In considering the economics of asteroid mining, the most favored candidates naturally belong to the class of near-Earth asteroids (NEAs)—those whose orbits bring them closest to Earth. The larger NEAs almost never approach the Earth to within the moon's proximity, but they do come relatively close, and they provide one significant advantage over lunar exploitation: their low surface gravity. For interplanetary exploration and exploitation, the crucial parameter that determines how much energy must be expended in order to move a ton of material describes the velocity needed to escape from an object's surface. The distance covered by the journey hardly matters, so long as the travel time does not grow too large.

One of the NEAs often mentioned as a potential target, 4660 Nereus (all asteroids have a number, and many have a name as well), has an elongated shape, with dimensions of about $500 \times 320 \times 320$ meters.[5] Nereus apparently consists largely of enstatite, a mineral made of magnesium, silicon, and oxygen; other, less abundant compounds on Nereus may include considerably rarer elements.[6] Silicon and oxygen, which are common on all solid objects in the solar system, form the bulk of rocks on Earth, but magnesium has a lower abundance, about one-twelfth of silicon's.[7] Nereus's orbit often brings it within a few million kilometers of Earth, and on some occasions, as will occur in 2060, its orbit will bring it to within 1.2 million kilometers—three times the moon's distance. On the moon's surface, the force of gravity equals one-sixth of Earth's, but Nereus's gravitational force, less than $1/3{,}000$ of the moon's, would allow machines, or even individual miners, to fling material from the asteroid's surface into space with modest expenditures of energy. (The gravitational force that an object exerts at its surface varies in proportion to the object's radius times its density. If the moon and Nereus have about the same density, the moon's gravitational force will exceed Nereus's by a factor of about 3,500.)

Even an asteroid as small as Nereus could provide a rich mineral lode analogous to an ore body on Earth measuring hundreds of meters in each dimension. The richest ore body in North America, found near Timmins, Ontario, in 1963, contained about 30 million tons of desirable minerals, including more than 12,000 tons of silver.[8] Nereus has a comparable mass ripe for appropriation—along with the obvious issues of getting and taking that require serious examination.

If one wants to think bigger, consider the asteroid Psyche, which orbits the sun in the heart of the asteroid belt, where it always remains at least 150 million kilometers farther from us than the planet Mars. Psyche consists mainly of iron, nickel, and associated elements and measures nearly 200 kilometers across—the largest such object in the solar system, and perhaps the would-be core of a planet that never formed. NASA plans to send a probe toward Psyche in 2022, with arrival scheduled four years later at an object holding 20 million times the raw materials that the much smaller, much closer Nereus contains—or roughly a million times more than all the metal mined from Earth.[9]

Before we leap one or two decades into the future, if only in imagination, we might pause to examine any international agreements that deal with mining the moon and the asteroids. An initial, salient point about these agreements, which may also prove relevant to the future colonization of exoplanets, lurks within the fact that most readers have little or no knowledge of any such agreements. This testifies to the agreements' potential impotence, which will probably increase once actual mining operations within the solar system become more promising.

The future of space law pivots primarily on the United Nations' Outer Space Treaty, known more formally as the "Treaty on Principles Governing the Activities of States in the Exploration and Use of Outer Space, including the Moon and Other Celestial Bodies."[10] First made permanent in 1959, the treaty has now received ratification from more than 100 nations, including all those likely to be involved in space exploration and space exploitation, except Iran. From an earthly point of view, the treaty's key provisions forbid the placement of nuclear or other weapons of mass destruction on the moon, in orbit, or anywhere else in outer space, and they state that all celestial bodies are to be used exclusively for peaceful purposes. From a more far-flung perspective, the treaty states that outer space, including all celestial bodies, is not subject to national appropriation by claim of sovereignty, by means of use or occupation, or by any other means, and that all parties to the treaty will follow international law in their activities relating to the exploration and use of outer space.

The phrase "national appropriation" either opens a loophole through which massive mining operations may proceed, or, by a more restrained reading, prevents any such operations from adding to a nation's wealth. This leaves the possibility—currently the only one under active investigation—that organizations not tied to any state can appropriate what they like and what they can from other objects in the solar system. In 2015, the United States Spurring Private Aerospace Competitiveness and Entrepreneurship (SPACE) Act went into force. The act denies that the United States asserts sovereignty over any cosmic objects, but it allows specifically for U.S. citizens (including corporations) to engage in the commercial exploration and exploitation of "space resources," which, however, do not include any biological life forms that may exist in space.[11]

Whether any appropriation of space resources would represent a good or a bad result provides an intriguing topic for a debate that no one seems interested in promoting. Toward the end of 2016, Deep Space Industries (DSI), a privately held American corporation, announced its plans to create Prospector-X, a probe designed to test the technologies required to achieve automated asteroid mining at a comparatively low cost.[12] Since Prospector-X will only orbit the Earth, asteroid mining can hardly lie in DSI's immediate future, but in the longer term, no technological obstacle seems to bar the arrival of an era when this sort of mining will become not only feasible but also common, with a host of competitors and, we may hope, only peaceful confrontations among corporations seeking to mine a particular asteroid. Luxembourg, where DSI has its headquarters, has shown an impressive eagerness to engage in this project.[13]

Unsurprisingly, DSI does not see the Outer Space Treaty as an obstacle to these activities. The company's attitude receives considerable support from the SPACE Act's relevant passage in Title IV of this legislation, which calls upon the President to

> facilitate commercial exploration for and commercial recovery of space resources by United States citizens; discourage government barriers to the development of economically viable, safe, and stable industries for commercial exploration for and commercial recovery of space resources in manners consistent with

the international obligations of the United States; and promote the right of United States citizens to engage in commercial exploration for and commercial recovery of space resources free from harmful interference, in accordance with the international obligations of the United States and subject to authorization and continuing supervision by the Federal Government.[14]

Perusal of this key passage allows the reader to admire the phrase "commercial recovery of space resources," which might cause a naïve reader to conclude that such resources have somehow been lost. However, the phrase presumably refers to the act of extracting minerals, as defined in 30 U.S. Code §1403—something like a football team "recovering" a fumble by taking the ball from the other team's possession. We should also note that the legislation carefully cites the United States' international obligations, which contain no explicit prohibition of the commercial use or extraction of space resources. After serious asteroid-mining operations have begun, legal arguments may increase over what activities may or may not conform to the Outer Space Treaty.[15] The most important issues, from another perspective, deal with the practicalities of who can actually succeed in extracting resources from celestial objects, and how they can do so while avoiding any claim to ownership or rivalry for "recovery" on a particular object. Sagi Kfir, DSI's general counsel and a pro-mining advocate, has compared space resources to fish in the sea, which are extracted without any claim to ownership of the water.[16] This analogy overlooks the facts that harvested minerals on asteroids, unlike fish, do not reproduce themselves; that many species of fish would disappear if governments did not regulate fishing; and that if one mines enough of an asteroid, there will be nothing left to claim.

The reader who has followed this discussion may well ask, What do these considerations imply about the future of human activities that occur not within the solar system, but instead on distant exoplanets? Whatever happens within the solar system will likely reverberate through future attempts to exploit distant planetary realms by colonization or other means. For now, this extrapolation qualifies as speculation, even hyperbole. Any reasonable prediction of the uncertain future of humanity places the colonization of exoplanets at least a few centuries in our future, when (so dreams are made) spacecraft carrying would-be colonists on hundred-light-year journeys toward other worlds will set out from what might be an Earth dying as the result of previous human activities, including the extraction and consumption of once-buried minerals.

To those who find these notions entirely fanciful, we cite the statement that Stephen Hawking made in 2016: "Although the chance of a disaster to planet Earth in a given year may be quite low, it adds up over time, and becomes a near certainty in the next thousand or ten thousand years. By that time we should have spread out into space, and to other stars, so a disaster on Earth would not mean the end of the human race."[17] Looking toward the comparatively near future, Hawking stated that if humanity hopes to avoid extinction, we must plan to become a multiplanet species within the next century, presumably by colonizing Mars.[18] To this apocalyptic vision, the physicist Freeman Dyson offers a single word of judgment: "rubbish!"[19] The author concurs in the view that if we cannot solve our problems here on Earth, outer space offers no likelier avenues to long-term success.

Exoplanets' hypothesized benefit as future abodes for humanity brings a more immediate boon: They allow us to plunge enjoyably into the deep end of the pool of speculation. While we are there, we might give a nod to the possibility that even before Hawking's grim future may become reality, we might possibly experience a great deal by interacting with exoplanetary civilizations that have preceded us down the broad highway of history and have survived—to enlighten or to eliminate us.

NOTES

Prologue

1. Gregory Laughlin and Jack Lissauer, "Exoplanetary Geophysics—An Emerging Discipline," in *Treatise on Geophysics, Second Edition,* ed. Gerald Schubert (Amsterdam: Elsevier Publishing, 2015), vol. 10, 673–694. The cited quotation appears on page 689.

2. Scott Gaudi, private communication, November 14, 2017.

3. Paul Butler, private communication, November 26, 2017.

4. Michel Mayor and Didier Queloz, "A Jupiter-Mass Companion to a Solar-Type Star," *Nature* 378 (1995): 355–359; Geoffrey Marcy and Paul Butler, "A Planetary Companion to 70 Virginis," *Astrophysical Journal Letters* 464 (1996): L147–151; Paul Butler and Geoffrey Marcy, "A Planet Orbiting 47 Uma," *Astrophysical Journal Letters* 464 (1996): L153–156. More generally, see Michel Mayor and Pierre-Yves Frei, *New Worlds in the Cosmos: The Discovery of Exoplanets* (Cambridge: Cambridge University Press, 2003).

5. Donald Goldsmith, *Worlds Unnumbered* (Sausalito, CA: University Science Books, 1997); Govert Schilling, *Tweeling Aarde—De Speurtocht Naar Leven in Andere Planetenstelsels* (Amsterdam: Wereldbibliotheek, 1997).

1. The Long Search for Other Solar Systems

1. Lucretius, *De Rerum Natura,* Book II, lines 1084–1086 (Cambridge, MA: Loeb Classical Library of Harvard University Press, 2nd ed., 1924).

2. Alexander Pope, *An Essay on Man. In Epistles to a Friend. Epistle I. Corrected by the Author,* line 21 (New York: Gale Ecco Print Editions, 2010).

3. For a general overview of these techniques, see Michael Lemonick, *Mirror Earth: The Search for Our Planet's Twin* (New York: Walker & Company, 2012).

2. Cosmic Distances

1. Donald Goldsmith, *The Runaway Universe: The Race to Discover the Future of the Universe* (New York: Basic Books, 2000).

2. David Clark and Matthew Clark, *Measuring the Cosmos: How Scientists Discovered the Dimensions of the Universe* (New Brunswick, NJ: Rutgers University Press, 2013).

3. William Waller, *The Milky Way: An Insider's Guide* (Princeton, NJ: Princeton University Press, 2013).

3. Early Quests for Exoplanets

1. Virginia Trimble, "The Quest for Other Worlds, 350 BCE to 1995 BCE," in *The Search for Other Worlds: Fourteenth Astrophysics Conference,* ed. Stephen Holt and Drake Deming (New York: American Institute of Physics, 2004).

2. Ray Jayawardhana, *Strange New Worlds: The Search for Alien Planets and Life beyond Our Solar System* (Princeton, NJ: Princeton University Press, 2011).

3. Edward Kolb, *Blind Watchers of the Sky: The People and the Ideas That Shaped Our Ideas of the Universe* (New York: Basic Books, 1996).

4. For a discussion of the astrometric and radial-velocity methods, see Donald Goldsmith, *Worlds Unnumbered* (Sausalito, CA: University Science Books, 1996), and Jayawardhana, *Strange New Worlds.*

5. Peter van de Kamp, "Astrometric Study of Barnard's Star from Plates Taken with the 24-inch Sproul Refractor," *Astronomical Journal* 68 (1963): 515–521; M. Kürster, "The Low-Level Radial Velocity Vari-

ability in Barnard's Star (= GJ 699): Secular Acceleration, Indications for Convective Redshift, and Planet Mass Limits," *Astronomy & Astrophysics* 403 (2003): 1077–1087.

6. For general information on the Gaia spacecraft, see the ESA website, http://sci.esa.int/gaia/.

7. Chadwick Trujillo and Scott Sheppard, "A Sedna-like Body with a Perihelion of 80 Astronomical Units," *Nature* 507 (2014): 471–474.

8. Michael Brown and Konstantin Batygin, "Evidence for a Distant Giant Planet in the Solar System," *Astronomical Journal* 151 (2016): 22.

9. Cory Shankman et al., "OSSOS VI: Striking Biases in the Detection of Large Semimajor Axis Trans-Neptunian Objects," *Astronomical Journal* 154 (2017): 50.

10. Konstantin Batygin interview, March 31, 2017.

11. Aleksander Wolszczan and Dale Frail, "A Planetary System around the Millisecond Pulsar 1257 + 12," *Nature* 355 (1992): 145–147.

12. For a discussion of neutron stars and pulsars, see Geoff McNamara, *Clocks in the Sky: The Story of Pulsars* (New York: Springer Praxis Books, 2008).

13. Stephen Thorsett et al., "The Triple Pulsar System PSR B1620–26 in M4," *Astrophysical Journal* 523 (1999): 763–770.

14. Donald Backer, Richard Foster, and Shauna Sallmen, "A Second Companion of the Millisecond Pulsar 1620–26," *Nature* 365 (1993): 817–819.

4. The Breakthrough: Measuring Radial Velocity Precisely

1. John Asher Johnson, *How Do You Find an Exoplanet?* (Princeton, NJ: Princeton University Press, 2016).

2. Bruce Campbell, Gordon Walker and Stephenson Yang, "A Search for Substellar Companions to Solar-Type Stars," *Astrophysical Journal* 331 (1999): 902–921; Gordon Walker et al., "Gamma Cephei— Rotation or Planetary Companion?," *Astrophysical Journal Letters* 396 (1992): L91–94; Artie Hatzes et al., "A Planetary Companion to γ Cephei A," *Astrophysical Journal* 599 (2003): 1383–1394.

3. Jacob Berkowitz, "Lost World: How Canada Missed Its Moment of Glory," *Globe and Mail,* September 25, 2009, available at

https://www.theglobeandmail.com/technology/science/lost -world-how-canada-missed-its-moment-of-glory/article4290133 /?page=all.

4. Dava Sobel, *A More Perfect Heaven: How Copernicus Revolutionized the Cosmos* (New York: Walker & Company, 2011); Konrad Rudnicki, "The Generalized Cosmological Copernican Principle," available at http://southerncrossreview.org/51/rudnicki4.htm.

5. Data available at http://exoplanet.eu/catalog/hd_20782_b/.

6. Ralph Waldo Emerson, *Compensation* (1841).

7. Data for Proxima Centauri b are available at http://exoplanet.eu /catalog/proxima_cen_b/.

8. Ravi Kumar Kopparapu, "A Revised Estimate of the Occurrence Rate of Terrestrial Planets in the Habitable Zones around Kepler M-Dwarfs," *Astrophysical Journal Letters* 767 (2013): L8.

9. Andrew Howard interview, March 31, 2017.

10. Debra Fischer interview, May 2, 2017.

5. Finding Exoplanets by Their Transits

1. For general information about the CoRoT spacecraft, see the ESA website at http://www.esa.int/Our_Activities/Space_Science /COROT.

2. Gregory Henry et al., "A Transiting '51-Peg-Like' Planet," *Astrophysical Journal Letters* 529 (2000): L41–44; David Charbonneau et al., "Detection of Planetary Transits across a Sun-like Star," *Astrophysical Journal Letters* 529 (2000): L45–48; Alan Boss, *The Crowded Universe: The Search for Living Planets* (New York: Basic Books, 2009).

3. David Charbonneau interview, May 4, 2017.

4. For information about the Kepler spacecraft, see the NASA website at https://www.nasa.gov/mission_pages/kepler/main/index.html.

5. Eric Petigura, Andrew Howard, and Geoffrey Marcy, "Prevalence of Earth-Size Planets Orbiting Sun-like Stars," *Proceedings of the National Academy of Sciences* 110 (2013): 19273–19278.

6. Adam Burrows, "Highlights in the Study of Exoplanet Atmospheres," *Nature* 513 (2014): 345–352.

7. "NASA's K2 Mission: The Kepler Space Telescope's Second Chance to Shine," available at https://www.nasa.gov/feature/ames/nasas -k2-mission-the-kepler-space-telescopes-second-chance-to-shine.

8. Courtney Dressing and David Charbonneau, "The Occurrence Rate of Small Planets around Small Stars," *Astrophysical Journal* 767 (2013): 95–114; Courtney Dressing and David Charbonneau, "The Occurrence of Potentially Habitable Planets Orbiting M Dwarfs Estimated from the Full Kepler Dataset and an Empirical Measurement of the Detection Sensitivity," *Astrophysical Journal* 807 (2015): 45.

6. Directly Observing Exoplanets

1. Wesley Traub and Ben Oppenheimer, "Direct Imaging of Exoplanets," in *Exoplanets,* ed. S. Seager (Tucson: University of Arizona Press, 2010).
2. Neil Reid and Suzanne Hawley, *New Light on Dark Stars: Red Dwarfs, Low Mass Stars, Brown Dwarfs,* 2nd ed. (New York: Springer Praxis Books, 2005).
3. Adam Burrows et al., "The Theory of Brown Dwarfs and Extrasolar Giant Planets," *Reviews of Modern Physics* 73 (2001): 719–766.
4. Koraljka Mužić et al., "The Low-Mass Content of the Massive Young Star Cluster RCW 38," *Monthly Notices of the Royal Astronomical Society* 471 (2017): 3699–3712.
5. Alan Boss, "Giant Planet Formation: Theories Meet Observations," in *Planet Formation: Theories, Observations, and Experiments,* ed. Hubert Klahr and Wolfgang Brandner (Cambridge: Cambridge University Press, 2006).
6. Leon Golub and Jay Pasachoff, *The Solar Corona,* 2nd ed. (Cambridge: Cambridge University Press, 2009).
7. Jacques Beckers, "Adaptive Optics for Astronomy: Principles, Performance, and Applications," in *Annual Review of Astronomy and Astrophysics* 31 (1993): 13–62.
8. Anne-Marie Lagrange et al., "A Probable Giant Planet Imaged in the β Pictoris Disk," *Astronomy and Astrophysics* 493 (2009): L21–25.
9. For the planet's parameters, see http://exoplanet.eu/catalog/beta _pic_b/.
10. Christian Marois et al., "Direct Imaging of Multiple Planets Orbiting the Star HR 8799," *Science* 322 (2008): 1348–1352.
11. Travis Barman et al., "Simultaneous Detection of Water, Methane, and Carbon Monoxide in the Atmosphere of Exoplanet HR8799b,"

Astrophysical Journal 804 (2015): 61; Patrick Ingraham et al., "Gemini Planet Imager Spectroscopy of the HR 8799 Planets c and d," *Astrophysical Journal Letters* 794 (2014): L15.

12. G. Chauvin et al., "Discovery of a Warm, Dusty Giant Planet Around HIP 65426," *Astronomy and Astrophysics* 605 (2017): L9–17.

7. Detecting Planets with Einstein's Lens

1. Albert Einstein, "Die Grundlage der Allgemeinen Relativitätstheorie," *Annals of Physics* 354 (1916): 769–822.

2. John Asher Johnson, *How Do You Find an Exoplanet?* (Princeton, NJ: Princeton University Press, 2016).

3. Scott Gaudi interview, May 3, 2017.

4. Calen Henderson statement at Kepler & K2 Science Conference IV, NASA / Ames Research Center, June 20, 2017; see also Calen Henderson et al., "Candidate Gravitational Microlensing Events for Future Direct Lens Imaging," *Astrophysical Journal* 794 (2014): 71.

5. J. P. Beaulieu et al., "Discovery of a Cool Planet of 5.5 Earth Masses through Gravitational Microlensing," *Nature* 439 (2006): 437–440.

6. Kailash Sahu et al., "Relativistic Deflection of Background Starlight Measures the Mass of a Nearby White Dwarf Star," *Science* 356 (2017): 1046–1050.

8. Two Minor Methods for Finding Exoplanets

1. Michael Brooks, *13 Things That Don't Make Sense: The Most Baffling Scientific Mysteries of Our Time* (New York: Random House, 2008).

2. G. R. Harp et al., "SETI Observations of Exoplanets with the Allen Telescope Array," *Astronomical Journal* 152 (2016): 181.

3. S. Charpinet et al., "A Compact System of Small Planets around a Former Red-Giant Star," *Nature* 480 (2011): 496–499.

4. Sujan Sengupta, "Cloudy Atmosphere of the Extra-solar Planet HD189733b: A Possible Explanation of the Detected B-band Polarization," *Astrophysical Journal Letters* 683 (2008): L195.

9. A Gallery of Strange New Planets

1. Tabitha Boyajian et al., "Planet Hunters X. KIC 8462852—Where's the Flux?," *Monthly Notices of the Royal Astronomical Society* 457 (2016): 3988–4004.

2. Brian Lacki, "The High Rate of the Boyajian's Star Anomaly as a Phenomenon," available at https://arxiv.org/pdf/1610.03219.pdf.

3. Jonathan Katz, "Can Dips of Boyajian's Star Be Explained by Circumsolar Rings?," *Monthly Notices of the Royal Society* 471 (2017): 3680–3685.

4. Kimberly Cartier and Jason Wright, "Have Aliens Built Huge Structures around Boyajian's Star?," *Scientific American* news posting, May 1, 2017, available at https://www.scientificamerican.com/article/have-aliens-built-huge-structures-around-boyajian-rsquo-s-star/#; see also Jason Wright et al., "The G Search for Extraterrestrial Civilizations with Large Energy Supplies: IV. The Signatures and Information Content of Transiting Megastructures," *Astrophysical Journal* 816 (2016): 17.

5. Gerry Harp et al., "Radio SETI Observations of the Anomalous Star KIC 8462852," *Astrophysical Journal* 825 (2016): 155.

6. Jason Wright and Steinn Sigurðsson, "Families of Plausible Solutions to the Puzzle of Boyajian's Star," *Astrophysical Journal Letters* 829 (2016): L3.

7. Wright et al., "The G Search for Extraterrestrial Civilizations," 17.

8. Xavier Dumusque et al., "The Kepler-10 Planetary System Revisited by HARPS-N: A Hot Rocky World and a Solid Neptune-Mass Planet," *Astrophysical Journal* 789 (2014): 154.

9. Laurence Doyle et al., "Kepler-16: A Transiting Circumbinary Planet," *Science* 333 (2011): 1601–1606.

10. Elena Popova and Ivan Shevchenko, "Kepler-16b: Safe in a Resonance Cell," *Astrophysical Journal* 769 (2013): 152.

11. William Borucki et al., "Kepler-22b: A 2.4 Earth-Radius Planet in the Habitable Zone of a Sun-like Star," *Astrophysical Journal* 745 (2012): 120.

12. Jason Wright, "Kepler-22b: A 2.4 Earth-Radius Planet in the Habitable Zone of a Sun-like Star," at AstroWright: Astronomy and Meta-astronomy by Jason Wright, December 6, 2011, available at

https://sites.psu.edu/astrowright/2011/12/06/kepler-22b-a-2 -4-earth-radius-planet-in-the-habitable-zone-of-a-sun-like-star/.

13. James Owen and Timothy Morton, "The Initial Physical Conditions of Kepler-36B and C," *Astrophysical Journal Letters* 819 (2016): L10.

14. Stuart Clark, *The Search for Earth's Twin* (London: Quercus Publishing, 2016); Michael Lemonick, *Mirror Earth: The Search for Our Planet's Twin* (New York: Walker & Co., 2012).

15. Jon Jenkins and David Ciardi, "Discovery and Validation of Kepler-452b: A 1.6 $R_\oplus$ Super Earth Exoplanet in the Habitable Zone of a G2 Star," *Astronomical Journal* 150 (2015): 56.

16. "NASA's Kepler Mission Discovers Bigger, Older Cousin to Earth," NASA Press Release 15–56, July 23, 2015.

17. David Charbonneau et al., "Detection of Planetary Transits across a Sun-like Star," *Astrophysical Journal Letters* 529 (2000): L45–48.

18. Mark Swain, Gautam Vasisht, and Giovanni Tinetti, "The Presence of Methane in the Atmosphere of an Extrasolar Planet," *Nature* 452 (2008): 329–331; Mark Swain et al., "Water, Methane, and Carbon Dioxide Present in the Dayside Spectrum of the Exoplanet HD 209458b," *Astrophysical Journal* 704 (2009): 1616.

19. John Southworth et al., "Detection of the Atmosphere of the 1.6 Earth-Mass Exoplanet GJ 1132b," *Astronomical Journal* 153 (2017): 191.

20. Scott Gaudi interview, May 3, 2017.

21. B. Scott Gaudi et al., "A Giant Planet Undergoing Extreme-Ultraviolet Irradiation by Its Hot Massive-Star Host," *Nature* 546 (2017): 514–518.

22. Drake Deming et al., "Infrared Eclipses of the Strongly Irradiated Planet WASP-33b, and Oscillations of Its Host Star," *Astrophysical Journal* 754 (2012): 106; Korey Haynes et al., "Spectroscopic Evidence for a Temperature Inversion in the Dayside Atmosphere of the Hot Jupiter WASP-33b," *Astrophysical Journal* 806 (2015): 146.

23. Jason Dittmann et al., "A Temperate Rocky Super-Earth Transiting a Nearby Cool Star," *Nature* 544 (2017): 333–336.

24. Jack Lissauer et al., "All Six Planets Known to Orbit Kepler-11 Have Low Densities," *Astrophysical Journal* 770 (2013): 131.

25. Luca Borsato et al., "TRADES: A New Software to Derive Orbital Parameters from Observed Transit Times and Radial Velocities. Revisiting Kepler-11 and Kepler-9," *Astronomy and Astrophysics* 571 (2014): A38.

26. Juan Cabrera et al., "The Planetary System to KIC 11442793: A Compact Analogue to the Solar System," *Astrophysical Journal* 781 (2014): 18; Alexandre Santerne et al., "SOPHIE Velocimetry of *Kepler* Transit Candidates XVII: The Physical Properties of Giant Exoplanets within 400 Days of Period," *Astronomy and Astrophysics* 587 (2016): A64.

27. NASA Press Release 17–098, December 14, 2017; Nicholas St. Fleur, "An 8th Planet Is Found Orbiting a Distant Star, with A.I.'s Help," *New York Times*, December 15, 2017.

28. Stephen Kane and Dawn Gelino, "On the Inclination and Habitability of the HD 10180 System," *Astrophysical Journal* 792 (2014): 111.

29. Steven Vogt et al., "Six Planets Orbiting HD 219134," *Astrophysical Journal* 814 (2015): 12; Michael Gillon et al., "Two Massive Rocky Planets Transiting a K-dwarf 6.5 Parsecs Away," *Nature Astronomy* 1 (2017): Article 0056.

30. For general information on the TRAPPIST survey, see the web page maintained by the University of Liège at http://www.trappist.ulg.ac .be/cms/c_3300885/en/trappist-portail.

31. Information about TRAPPIST-1 is available at http://simbad.u -strasbg.fr/simbad/sim-basic?Ident=2MASS+J23062928-0502285.

32. "Three Potentially Habitable Worlds Found Around Nearby Ultra-cool Dwarf Star," ESO Science Release 1615, May 2, 2016, available at https://www.eso.org/public/news/eso1615/.

33. Information about TRAPPIST-1 is available at http://simbad.u -strasbg.fr/simbad/sim-basic?Ident=2MASS+J23062928–0502285.

34. TRAPPIST-1's mass provided in Songhu Wang et al., "Updated Masses for the TRAPPIST-1 Planets," *Astrophysical Journal* (2017), available at https://arxiv.org/abs/1704.04290.

35. Martin Cohen, *In Darkness Born: The Story of Star Formation* (Cambridge: Cambridge University Press, 1988).

36. See https://phys.org/news/2017-06-kepler-taught-rocky-planets -common.html.

37. Carl Sagan, "On the Origin and Planetary Distribution of Life," *Radiation Research* 15 (1961): 174–192.

38. Svante Arrhenius, *Worlds in the Making: The Evolution of the Universe* (New York: Harper & Row, 1908).

39. General information on the TESS spacecraft is available at the NASA website, https://tess.gsfc.nasa.gov/.

40. General information on the CHEOPS spacecraft is available at the ESA website, https://www.cosmos.esa.int/web/cheops.

41. Information on the NGTS is available at the ESO website, http://www.eso.org/public/teles-instr/paranal-observatory/ngts/.

42. For general information on the Evryscope project, see Nicholas Law et al., "Evryscope Science: Exploring the Potential of All-Sky Gigapixel-Scale Telescopes," *Publications of the Astronomical Society of the Pacific* 127 (2015): 234–249.

10. What Have We Learned?

1. Erik Petigura, Andrew Howard, and Geoffrey Marcy, "Prevalence of Earth-Size Planets Orbiting Sun-like Stars," *Proceedings of the National Academy of Sciences* 110 (2013): 19273–19278.

2. Dimitar Sasselov, *The Life of Super-Earths* (New York: Basic Books, 2012).

3. Benjamin Fulton et al., "The California Kepler Survey: III. A Gap in the Radius Distribution of Small Planets," *Astronomical Journal* 154 (2017): 109.

4. David Kipping, private communication, November 10, 2017.

5. Erik Petigura et al., "The California Kepler Survey: I. A High-Resolution Survey of 1305 Kepler Stars Hosting Kepler Transiting Planets," *Astronomical Journal* 154 (2017): 107.

6. Alexander Pope, *An Essay on Man. In Epistles to a Friend. Epistle I. Corrected by the Author,* lines 17–28 (New York: Gale Ecco Print Editions, 2010).

11. How Planets Form with Their Stars

1. Audrey Bouvier and Meenakshi Wadhwa, "The Age of the Solar System Redefined by the Oldest Pb-Pb Age of a Meteoritic Inclusion," *Nature Geoscience* 3 (2010): 637–641.

2. John Lewis, *Worlds without End: The Exploration of Planets Known and Unknown* (Reading, MA: Perseus Books, 1998).

3. Sean Wahl et al., "Comparing Jupiter Interior Structure Models to Juno Gravity Measurements and the Role of a Dilute Core," Geophysical Research Letters 44 (2017): 4649–4659.

4. Jack Lissauer, telephone interview, July 7, 2017.

5. Veronique Greenwood and Cassandra Willyard, "The Man Whose Models Revealed a Possible Ninth Planet in Our Solar System," *Popular Science*, September / October 2016.

6. Michiel Lambrechts and Anders Johanson, "Forming the Cores of Giant Planets from the Radial Pebble Flux in Protoplanetary Disks," *Astronomy and Astrophysics* 572 (2014): A107–118.

7. Harold Levison, "Growing the Terrestrial Planets from the Gradual Accumulation of Submeter-sized Objects," *Proceedings of the National Academy of Sciences* 112 (2015): 14180–14185.

8. General information on the ALMA array is available at http://www.almaobservatory.org/en/home/.

9. A general discussion of planetary migration is available at http://www.earth.northwestern.edu/people/seth/351/genesis.pdf.

10. Renu Malhotra, "Migrating Planets," *Scientific American* 281, no. 3 (September 1999): 56–63.

11. A general discussion of tidal locking is available at https://arxiv.org/pdf/1405.1025.pdf.

12. Michael Zhang and Kaloyan Penev, "Stars Get Dizzy after Lunch," *Astrophysical Journal* 787 (2014): 131.

13. Gijs Mulders et al., "A Super-solar Metallicity for Stars with Hot Rocky Exoplanets," *Astronomical Journal* 152 (2016): 187.

14. Bruce Macintosh, private communication, November 12, 2017.

15. Jason Wright, private communication, November 13, 2017.

16. Eric Agol, private communication, November 13, 2017.

12. Habitable Planets and the Search for Life

1. Abraham Loeb interview, May 4, 2017; Dimitar Sasselov interview, May 4, 2017; Guyon statement at Breakthrough Discuss Conference, Stanford University, April 20, 2017.

2. Su-Shu Huang, "The Problem of Life in the Universe and the Mode of Star Formation," *Publications of the Astronomical Society of the Pacific* 71 (1959): 421–424.

3. NASA / JPL Press Release, November 20, 2017, available at https://www.jpl.nasa.gov/news/news.php?release=2017-299.

4. Hisashi Hayakawa et al., "Long-Lasting Extreme Magnetic Storm Activities in 1770 Found in Historical Documents," *Astrophysical Journal Letters* 850 (2017): L31.

5. C. Robert Clauer and George Siscoe, "The Great Historical Geomagnetic Storm of 1859: A Modern Look," *Advances in Space Research* 38 (2006): 117–118.

6. Donald Goldsmith, "The $2 Trillion Economic Risk You Haven't Heard About," PBS online news story available at https://www.pbs.org/newshour/nation/the-2-trillion-economic-risk-you-havent-heard-about.

7. Manasvi Lingam and Abraham Loeb, "Risks for Life on Habitable Planets from Superflares of Their Host Stars," available at https://arxiv.org/pdf/1708.04241.pdf.

8. Manasvi Lingam and Abraham Loeb, "Impact and Mitigation Strategy for Future Solar Flares," available at https://arxiv.org/pdf/1709.05348.pdf.

9. Rachel Osten et al., "DRAFTS: A Deep, Rapid Archival Flare Transient Surge in the Galactic Bulge," *Astrophysical Journal* 754 (2012): 4.

10. Chuanfei Dong et al., "The Dehydration of Water Worlds via Atmospheric Losses," *Astrophysical Journal Letters* 847 (2017): 4; Chuanfei Dong et al., "Is Proxima Centauri b Habitable? A Study of Atmospheric Loss," *Astrophysical Journal Letters* 837 (2017): 26.

11. Xavier Bonfils et al., "A Temperate Exo-Earth Around a Quiet M Dwarf at 3.4 Parsecs," *Astronomy and Astrophysics* 607 (2017), available at https://www.researchgate.net/publication/321083637_A_temperate_exo-Earth_around_a_quiet_M_dwarf_at_34_parsecs.

12. Hinrich Schaefer et al., "A 21st-Century Shift from Fossil-Fuel to Biogenic Methane Emissions Indicated by $^{13}CH_4$," *Science* 352 (2016): 80–84.

13. Alexandra Witze, "Exoplanet Hunters Rethink Search for Alien Life," *Nature* 551 (2017): 421–422.

14. Shawn Domegal-Goldman interview, May 12, 2017.

13. Future Approaches to Hunting Exoplanets

1. In 1772, the French astronomer and mathematician Joseph-Louis Lagrange showed that if a less massive object such as a planet orbits a much more massive one such as a star, five specific locations, now called Lagrange points, exist in nearby space at which a third object can continue to follow a stable orbit without slowly drifting into an

entirely different position. Two of the five Lagrange points, denoted as L4 and L5, have positions along the planet's orbit, preceding or trailing the planet by one-sixth of the orbit's circumference, thus always equally distant from the star and the planet. The L3 point lies precisely on the opposite side of the star—not a good location for a spacecraft that has to send data to the planet. The L1 and L2 points lie along the line joining the star and planet. In the case of the sun and Earth, L1 lies 1.5 million kilometers closer to the sun than the Earth, and L2 is 1.5 million kilometers distant in the opposite direction. (Lagrange showed that, in the general case, both L1 and L2 will have distances from the planet equal to a particular fraction of the planet–star distance: the cube root of the ratio of the planet's mass to three times the star's mass!) Because any signals from the L1 point must compete with the sun's output, the L2 point offers the finest location for a spacecraft meant to stay far from Earth, but always in the same location with respect to our planet's orbit around the sun.

2. General information about the JWST is available at the NASA website, https://www.jwst.nasa.gov/.

3. General information about the Gaia spacecraft is available at the ESA website, http://sci.esa.int/gaia/.

4. For general information about asteroseismology, see Gerard Handler, "Asteroseismology," available at https://arxiv.org/pdf/1205.6407 .pdf

5. General information about PLATO is available at http://sci.esa.int /plato/.

6. General information about WFIRST and its heritage architecture is available at https://wfirst.gsfc.nasa.gov/observatory.html.

7. *New Worlds, New Horizons in Astrophysics* (Washington, DC: National Academies Press, 2010).

8. Lyman Spitzer, "The Beginnings and Future of Space Astronomy," *American Scientist* 50 (1962): 473–484.

9. General information about the WFIRST starshade is available at https://science.nasa.gov/technology/technology-stories /starshade-enable-first-images-earth-sized-exoplanets.

10. Chris Jones, "The Woman Who Might Find Us Another Earth," *New York Times Magazine,* December 7, 2016.

11. Sara Seager telephone interview, May 26, 2017.

12. General information about LUVOIR is available at https://asd.gsfc .nasa.gov/luvoir/.

13. General information about HabEx is available at https://www.jpl .nasa.gov/habex/.

14. "Jerry Nelson, Designer of the Segmented Telescope, Dies at 73," *New York Times,* June 21, 2017.

15. General information on the Keck Telescopes and the Gran Telescopio Canarias is available at http://www.keckobservatory.org/and http://www.gtc.iac.es/.

16. Information on SALT is available at https://www.salt.ac.za/.

17. Information on the GMT is available at https://www.gmto.org/.

18. Information on the EELT is available at https://www.eso.org/sci /facilities/eelt/.

19. General information about the TMT is available at http://www.tmt .org/.

20. Dennis Overbye, "Under Hawaii's Starriest Skies, A Fight over Sacred Ground," *New York Times,* October 3, 2016; Dennis Overbye, "Giant Telescope atop Hawaii's Mauna Kea Should Be Approved, Judge Says," *New York Times,* July 27, 2017.

21. If we divide 400 nanometers by 40 meters, the quotient provides an angular resolution of 10 one-billionths of a radian. Since a radian equals 57.296 degrees of arc, it contains 3,600 times 57.296, or 206,265, seconds of arc. Multiplication of 10 one-billionths by 206,265 provides the theoretical best resolving power: an ability to detect individual sources separated by more than 0.00206 seconds of arc.

22. General information on the VLA is available at http://www.vla.nrao .edu/.

23. Information on the NGVLA is available at https://science.nrao.edu /futures/ngvla.

24. Modern astrophysics teems with acronyms, especially in the realms of space agencies such as NASA and ESA—so much so that NASA appears to have assigned the same acronym, FINESSE, to two different future projects: the Fast INfrared Exoplanet Spectroscopy Survey Explorer *and* Field INvestigations to Enable Solar System Science and Exploration. This book includes references to organizations such as ESA, ESO, and NASA; space-based observatories that include CHEOPS, CoRoT, Gaia, HabEx, HST, JWST,

LUVOIR, PLATO, SIM, TESS, TPF, and WFIRST; and the Earth-bound telescopes and telescope systems known as CKS, EELT, ESPRESSO, EXPRES, GMT, GPI, HARPS, HIRES, MASCARA, MEarth, MOA, OGLE, SPHERE, TMT, and TRAPPIST. We may, perhaps, regard this profusion of acronyms as a tribute to our ability to explore the cosmos.

25. For information about SIM, see Stephen Unwin et al., "Taking the Measure of the Universe: Precision Astrometry with SIM Planet-Quest," *Publications of the Astronomical Society of the Pacific* 120 (2008): 38–88; for information about TPF, see https://science.nasa.gov/missions/tpf.

26. Slava Turyshev and Michael Shao, "Using the Sun as a Cosmic Telescope," *Scientific American* "Observations" (blog), 29 May 2017, available at https://blogs.scientificamerican.com/observations/using-the-sun-as-a-cosmic-telescope/#.

14. Proxima Calls: Can We Visit?

1. Information on the speed of the New Horizons spacecraft is available at https://en.wikipedia.org/wiki/New_Horizons.

2. Phil Lubin interview, March 30, 2017.

3. "Private Mission May Get Us Back to Enceladus Sooner than NASA," *New Scientist*, November 22, 2017, available at https://www.newscientist.com/article/mg23631533-900-private-mission-may-get-us-back-to-enceladus-sooner-than-nasa/.

4. Albert Einstein, "Zur Elektrodynamik bewegter Körper," *Annalen der Physik* 17 (1905): 891–921.

5. Marina Brozovic et al., "Radar Observations and a Physical Model of Asteroid 4660 Nereus, a Prime Space Mission Target," *Icarus* 201 (2009): 153–166.

6. "Enstatite mineral data," available at http://www.webmineral.com/data/Enstatite.shtml#.WXo5saIrLBI.

7. See, generally, Thomas Hamilton, *Dwarf Planets and Asteroids: Minor Bodies of the Solar System* (Houston, TX: Strategic Book Publishing & Rights Agency, LLC, 2014).

8. Len Gillis, "Timmins Mine Likely to Close by 2022," *News Provincial* article, November 17, 2016, available at http://www.thesudburystar.com/2016/11/17/timmins-mine-likely-to-close-by-2022.

9. Information on NASA's mission to Psyche is available at https://www
.nasa.gov/feature/jpl/nasa-moves-up-launch-of-psyche-mission-to
-a-metal-asteroid.

10. The text of the Outer Space Law treaty is available at http://www
.unoosa.org/oosa/en/ourwork/spacelaw/treaties/outerspacetreaty
.html.

11. See Maggie Koerth-Baker, "Who Makes the Rules for Outer Space?,"
PBS / NOVA posting, November 30, 2015, available at http://www
.pbs.org/wgbh/nova/next/space/space-law/; the text of the SPACE
act is available at https://www.congress.gov/bill/114th-congress
/house-bill/2262.

12. Alex Létourneau, "Asteroid Mining Becoming More of a Reality,"
Forbes, January 25, 2013.

13. Mariella Moon, "Luxembourg's Asteroid Mining Law Takes Effect
August 1st [2017]," available at https://www.engadget.com/2017
/07/30/luxembourg-asteroid-mining-law-august-1/.

14. Title IV, Section (402) of the SPACE Act, available at https://www
.congress.gov/bill/114th-congress/house-bill/2262.

15. Kenneth Chang, "If No One Owns the Moon, Can Anyone Make
Money Up There?," *New York Times,* November 26, 2017.

16. Sagi Kfir, "Is Asteroid Mining Legal? The Truth Behind Title IV of
the Commercial Space Launch Competitiveness Act of 2015," available at http://deepspaceindustries.com/is-asteroid-mining-legal/.

17. Peter Holley, "Stephen Hawking Just Gave Humanity a Due Date
for Finding Another Planet," *Washington Post,* November 17, 2016.

18. Sara Fecht, "Stephen Hawking Says We Have 100 Years to Colonize
a New Planet—Or Die. Can We Do It?," *Popular Science,* May 4,
2017, available at http://www.popsci.com/stephen-hawking-human
-extinction-colonize-mars.

19. Freeman Dyson interview, May 8, 2017.

FURTHER READING

Batygin, Konstantin, Gregory Laughlin, and Alessandro Morbidelli. "Born of Chaos: New Evidence Suggests the Solar System's Early Eras Were Defined by Wandering Worlds and Staggering Displays of Interplanetary Destruction." *Scientific American* 314, no. 5 (May 2016): 28–37.

Boss, Alan. *The Crowded Universe: The Search for Living Planets.* New York: Basic Books / Perseus Books, 2009.

Clark, Stuart. *The Search for Earth's Twin.* London: Quercus Publishing, 2016.

Cohen, Nathan. *Gravity's Lens.* New York: John Wiley and Sons, 1989.

Ellerbroek, Lucas. *Exoplanets: The Search for Extraterrestrial Life.* London: Reaktion, 2017.

Finkbeiner, Ann. "Near Light-Speed Mission to Alpha Centauri." *Scientific American* 316, no. 3 (March 2017): 30–37.

Goldsmith, Donald. *The Runaway Universe: The Race to Discover the Future of the Universe.* New York: Basic Books, 2000.

———. *Worlds Unnumbered.* Sausalito, CA: University Science Books, 1997.

Goldsmith, Donald, and Tobias Owen. *The Search for Life in the Universe.* 3rd ed. Sausalito, CA: University Science Books, 1997.

Hall, Shannon. "The Secrets of Super-Earths." *Sky & Telescope* 133, no. 3 (March 2017): 22–29.

Jayawardhana, Jay. *Strange New Worlds: The Search for Alien Planets and Life beyond Our Solar System*. Princeton, NJ: Princeton University Press, 2011.

Johnson, John Asher. *How Do You Find an Exoplanet?* Princeton, NJ: Princeton University Press, 2016.

Kitchin, Chris. *Exoplanets: Finding, Exploring, and Understanding Alien Worlds*. New York: Springer, 2011.

Lemonick, Michael. *Mirror Earth: The Search for Our Planet's Twin*. New York: Walker & Co., 2012.

———. "The Search for Planet X." *Scientific American* 314, no. 2 (February 2016): 30–37.

Lewis, John S. *Worlds without End: The Exploration of Planets Known and Unknown*. Reading, MA: Helix Books / Perseus Books, 1998.

Mayor, Michel, and Pierre-Yves Frei. *New Worlds in the Cosmos: The Discovery of Exoplanets*. Cambridge: Cambridge University Press, 2003.

Mermin, N. David. *It's About Time: Understanding Einstein's Relativity*. Princeton, NJ: Princeton University Press, 2009.

Perryman, Michael. *The Exoplanet Handbook*. Cambridge: Cambridge University Press, 2014.

Sasselov, Dimitar. *The Life of Super-Earths: How the Hunt for Alien Worlds and Artificial Cells Will Revolutionize Life on Our Planet*. New York: Basic Books / Perseus Books, 2012.

Schilling, Govert. *Tweeling Aarde—De Speurtocht Naar Leven in Andere Planetenstelsels*. Amsterdam: Wereldbibliotheek, 1997.

Stevenson, David. *Under a Crimson Sun: Prospects for Life in a Red Dwarf System*. New York: Springer, 2013.

Tasker, Elizabeth. *The Planet Factory*. New York: Bloomsbury, 2017.

Websites

The European Exoplanet Catalog. http://exoplanet.eu/catalog/.

Exoplanet Data Explorer. http://exoplanets.org/table?datasets=other.

Exoplanet Discoveries: Latest Data from NASA's Exoplanet Archive. https://exoplanets.nasa.gov/discoveries-overlay/.

Habitable Exoplanets Catalog. http://phl.upr.edu/projects/habitable -exoplanets-catalog.

NASA / Caltech Exoplanet Archive. https://exoplanetarchive.ipac.caltech .edu/.

ACKNOWLEDGMENTS

In writing this latest book, I have once again benefited greatly from conversations with a large number of astronomers who cheerfully provided me with information and scientifically informed opinions about the topics mentioned in this work. In addition, many of those who spoke at astronomical conferences willingly answered my questions, adding to the enlightenment that I received from their presentations. Some of these astronomers most kindly adapted diagrams that appeared in their published work for publication in this book, as I have acknowledged in the figure captions.

I want to thank Eric Agol, Bruce Balick, Gibor Basri, Konstantin Batygin, Jean-Phillipe Beaulieu, Howard Bond, Alan Boss, Kenneth Brecher, Adam Burrows, Butler Burton, Paul Butler, David Catling, Catherine Cesarsky, David Charbonneau, Eugene Chiang, Drake Deming, George Djorgovski, Courtney Dressing, Freeman Dyson, Daniel Fabrycky, Henry Ferguson, Debra Fischer, Benjamin Fulton, Scott Gaudi, Benjamin Gerard, Daniel Gezari, Suvi Gezari, Michel Gillon, Shawn Domegal-Goldman,

Paul Goldsmith, Todd Henry, Matthew Holman, Andrew Howard, Jeremy Kasdin, Steve Kilston, David Kipping, Robert Kirshner, Heather Knutson, Lawrence Krauss, Edward Krupp, Marc Kuchner, David Latham, Hal Levison, Nikole Lewis, Jack Lissauer, Avi Loeb, Philip Lubin, Bruce Macintosh, Stephen Maran, Lawrence Marschall, Dimitri Mawet, Tsvi Mazeh, Victoria Meadows, Erik Petigura, Mark Postman, Fred Rasio, Martin Rees, Jason Rowe, Charli Sakari, Kailash Sasu, Dimitar Sasselov, Sara Seager, Ji-Ming Shi, David Spergel, Amiel Sternberg, Edward Stone, Jill Tarter, Virginia Trimble, David Werner, Joshua Winn, Jason Wright, Jennifer Yee, and most particularly Charles Beichman for providing information and suggestions that helped me in writing this book. Cody Andresen and Laura Haertling of Studio Percolate Design and Photography expertly created new diagrams. At Harvard University Press, my gratitude goes to Esther Blanco Benmaman, Susan Wallace Boehmer, Emeralde Jensen-Roberts, Stephanie Vyce, and most notably to my editor, Jeff Dean. At Westchester Publishing Services, Mary Ribesky and Helen Wheeler were extremely helpful. In addition, I am grateful to the anonymous reviewers of this manuscript who provided many significant suggestions and corrections.

INDEX